MANAGEMENT AND ENGINEERING OF
FIRE SAFETY AND LOSS PREVENTION

Papers presented at the 3rd International Conference on Management and Engineering of Fire Safety and Loss Prevention, 18–20 February 1991, Aberdeen.

MANAGEMENT AND ENGINEERING OF FIRE SAFETY AND LOSS PREVENTION

ONSHORE AND OFFSHORE

Edited by

BHR GROUP LTD

Cranfield, Bedford, UK

Taylor & Francis Group

LONDON AND NEW YORK

Organized by BHR Group Limited

Published 2005 by Taylor & Francis
2 Park Square, Milton Park, Abingdon, Oxon, OX14 4RN
52 Vanderbilt Avenue, New York, NY 10017, USA

First issued in paperback 2020

*Taylor & Francis is an imprint of the Taylor & Francis Group,
an informa business*

Copyright © 2005, Taylor & Francis.

British Library Cataloguing in Publication Data

International Conference on Management and Engineering of
Fire Safety and Loss Prevention (3rd: 1991: Aberdeen,
Scotland)
    Management and engineering of fire safety and loss
    prevention: onshore and offshore.
    I. Title   II. BHR Group
    658.382

ISBN 13: 978-0-367-58005-6 (pbk)
ISBN 13: 978-1-85166-676-8 (hbk)

Library of Congress CIP data applied for

CONTENTS

SESSION B: RESEARCH, RISK REDUCTION AND DESIGN SAFETY

SESSION C: DETECTION AND CONTROL

SESSION D: PROTECTIVE SYSTEMS

ACKNOWLEDGEMENTS

The valuable assistance of the Technical Advisory Committee and panel of referees is gratefully acknowledged.

Technical Advisory Committee

Mr C. P. A. Thompson (Chairman)	Bechtel Ltd
Mr R. Bell	Health and Safety Executive
Mr M. Broadribb	BP Petroleum Development Ltd
Dr E. J. Denney	Loss Prevention Council
Dr H. Hughes	UK Offshore Operators Association Ltd
Mr R. Mayson	BNFL plc
Mr H. Richardson	A & H Associates Ltd
Dr L. Small	Cremer & Warner
Mr R. A. Whiteley	Wormald Engineering

Overseas Corresponding Members

Mr B. Bang	Danish Energy Agency, Denmark
Mr P. Lund	Society of Fire Protection Engineers, USA
Mr E. J. Thomas	Phillips Petroleum Co., USA
Mr R. Vondrasek	National Fire Prevention Association, USA

SESSION A:RISK ASSESSMENT

QUANTITATIVE RISK ASSESSMENT: LIKELY ERROR RANGES AND THE NEED FOR A MULTI DISCIPLINARY APPROACH

Martin McD. Grant (WS Atkins Engineering Sciences, Aberdeen)

ABSTRACT

The importance of being able to model the consequences of hydrocarbon accident events offshore is discussed. Using the specific example of structural collapse caused by hydrocarbon fire, an attempt is made to estimate the magnitude of errors typically associated with risk analyses. Those areas which, on the basis of current knowledge, are likely to generate the largest errors are identified. The reduction of such errors is seen to depend on the application of expertise which spans a number of disciplines. Finally the manner in which a risk analysis should be used is discussed given the errors that will typically be present.

1. INTRODUCTION

The early development of quantified risk analysis took place largely in the aerospace and nuclear industries. In these industries there is a readily definable 'top event' (airplane crash, reactor meltdown) whose consequences are regarded as being wholly unacceptable. For this reason the safety case tends to be based heavily on probabilistic arguments; that is to say it must be demonstrated that the 'top event' has only a certain (low) probability of occurring. The safety case does not tend to rely on demonstrating that the consequences of the top event are in some way tolerable.

The situation in the process industries is different with the application of quantified risk assessment having a greater emphasis on the examination of incident consequences. That this is so is related to the fact that it is not economic, nor is it necessary from a safety viewpoint, to construct process plant such that leakage of hazardous product has a 'negligible' probability. For an offshore installation it is necessary to demonstrate that process leakage cannot by itself or via an escalation of the incident lead to significant loss of life or platform loss. For small onshore chemical plants it may simply be necessary to show that such leakage cannot generate a hazard to the surrounding population.

Thus the process risk assessment will typically consist of two elements as follows:

> probabilistic analysis involving definition of component failure rates and their manipulation by fault and event tree analysis

consequence analysis involving the study of releases of flammable and toxic materials in order to determine their effect on plant and personnel. Particular attention is usually given to studying the manner in which initiating events may escalate to cause catastrophic events

The results of these two studies will typically be combined in order to generate a risk measure expressed as potential fatalities per annum or some other such indicator.

2. CONSEQUENCE MODELLING

The strong emphasis on the analysis of consequences as part of risk analysis in the process industries necessitates consideration of a very broad range of issues.

In particular, the process risk analyst must be able to model the following types of phenomena·
- release rates of liquids and gases from vessel and pipeline breaches including two phase effects

- dispersion of released liquids and gases

- toxic effects on personnel

- heat radiation from jet, pool and flash fires

- response of structures to heat loading

- effects of heat radiation on personnel

- overpressures from confined, semi confined and unconfined explosions

- effects of overpressure on structures

- effects of overpressure on personnel

Given the wide range of phenomena to be considered and their complexity, two problems present themselves:

- the present day understanding of these phenomena is not always complete

- there are gaps in the knowledge base. In the offshore industry a specific example is the prediction of blast overpressures from partially confined explosions;

- problems span a number of disciplines - e.g. modelling the effects of a gas explosion will require consideration of fluid dynamics, combustion processes and structural response. It is unrealistic to expect one individual to cover each aspect of such a problem.

The author believes that these difficulties are not always appreciated. In particular, it seems that high accuracy in certain aspects of the modelling process is pursued without an appreciation of the gross errors associated with other facets. Put another way, accuracy in one element of the calculation process is used to yield false security regarding the validity of the overall result.

It is outwith the scope of this paper to detail the uncertainties associated with all aspects that must be considered when modelling the consequences of process related accidents. As an alternative, the uncertainties associated with the modelling of one particular phenomenon are examined. The particular example chosen is that of structural response to hydrocarbon fires. This example is chosen in particular as the author believes it has often been analysed in a non- consistent manner. Specifically there has been a tendency to over simplify certain aspects of the problem whilst concentrating on those that are more tractable.

3. STRUCTURAL RESPONSE TO HYDROCARBON FIRES

The manner in which structures respond to fires is critical to the safety of offshore installations. Structural collapse may lead to the following:

- involvement of further process inventory thereby leading to incident escalation

- loss of safety systems

- loss of escape routes

- loss of safe havens

- ultimately, loss of the entire installation

It is therefore important to be able to predict the time to structural collapse. This information would be used to identify whether structural integrity is maintained throughout the incident or at least long enough to permit safe evacuation.

The various steps in this calculation process and their associated uncertainties are discussed below. The discussion covers both pool and jet fires. It should be remembered that the following is not intended to be a comprehensive review of the literature on this subject. Rather, it is intended to be an overview which has the aim of identifying the primary sources and likely magnitudes of error in such an analysis.

3.1 Outflow rate (jet fires)

For a gaseous jet fire, the amount of fuel feeding a fire is simply the outflow rate. There should be no large error in calculating this parameter although there may be some uncertainty in estimating the orifice loss coefficient. It should however, always be possible to keep this error on the side of conservatism.

For two phase releases, the situation is much more complicated for a number of reasons. Firstly there is the difficulty of calculating the outflow rate especially if, in the case of releases from pipelines, the two phase regime extends back along the pipe. In this case the vapour component causes a reduction in average density thereby reducing the flowrate when compared to the liquid only case. There are further difficulties associated with determining whether a stable jet fire will be formed; this will be dependent on the liquid to vapour ratio. Another problem area is whether the heat from the fire will significantly increases the vaporisation fraction. The heat may arise from the jet fire itself or a pool fire arising from the fall out of liquids.

The state of knowledge on these factors is limited and it is difficult to estimate the error ranges introduced when trying to model jet fires from two phase releases. However it seems reasonable to assume that they are potentially large.

3.2 Burning rate (pool fires)

The controlling parameter concerning the amount of fuel feeding a pool fire is the burning rate. Babrauskas (1983) presents burning rate data for 22 materials together with their associated uncertainty ranges. The uncertainty ranges rarely exceed +/-10%. There is a wide range of burning rates quoted for crude oil but this is probably due to the differing compositions used in different tests.

Where experimentally derived burning rate data are not available it is necessary to calculate this parameter. Such an approach is most likely to be necessary when the liquid is a mixture. Mudan (1984) reviews a number of equations for calculating the burning rates of mixture. They are found to calculate mass burning rate well (generally within 20%) even where the components have widely differing boiling points.

Burning rate will differ depending on whether the release is on water or on land. However, there is a significant amount of data regarding this aspect on which to base a judgement. For example, Mudan (1984) notes that the burning rate of LPG increases by a factor of two if on water as opposed to land. However,the present author is aware of no literature which covers the effect of waves on the burning rates of releases on water.

The effect of wind on burning rate is covered in section 3.5. Diameter is known to have an influence on burning rate but this seems to be confined to pools of less than 1 metre diameter (Mudan, 1984 and Babrauskas, 1983): process accidents are likely to lead to larger pools than this so it seems that the diameter effect will not be significant.

In summary it appears that, in general, the calculation of burning rate is well understood and is unlikely to introduce errors of more than +/- 20% to the calculation.

It should be noted that if estimates of Surface Emissive Powers (SEP) are available then it is not necessary to calculate burning rate as this is just one step towards calculating the SEP. The errors associated with the use of direct SEP measurements are discussed in section 3.6.

3.3 Flame Model

Early efforts at modelling the radiation from fires treated
the flame as a point source emitter. However, this approach cannot be used to accurately predict radiation levels in the near field (say within 2-3 flame lengths). Errors generated by use of a point source model may be over or underestimates depending on the position being considered.

An improvement on the point source method is the multiple point source method whereby the flame is split into a series of point sources along its axis. However more recent work (e.g. Chamberlain, 1987) has demonstrated that it is better to consider the flame as a solid body which emits radiation over its entire surface.

Crocker and Napier (1988) have compared the results of the three types of model discussed above and some of their results are reproduced below. They relate to a 0.08m diameter release of LPG with a driving pressure of 19.3 Bar.

MODEL	DISTANCE TO RADIATION LEVEL (KW/m^{-2})		
	12.6	4.7	1.6
Point Source	65	111	189
Multiple Point Source	166	306	596
Surface Emitter	135	201	318

Significant discrepancy is apparent between the models. It should be noted that, as stated in the paper, the reasons for the discrepancies include the manner in which wind effects are taken into account as well as the type of flame model used. However, additional information available to the present author supports the fact that a factor of two error is typically introduced by using a point source as opposed to surface emitter model.

Although the surface emitter model is generally regarded as giving accurate results it should be noted that in fact flames emit from their entire volume. The present author is aware of no model which takes this factor directly into account although it could be argued that the problem is at least partially addressed by the combined multiple point source/surface emitter model of McMurray (1982).

3.4 Flame shape

For any radiation modelling exercise it will be necessary to estimate the flame shape. If using a surface emitter model the following must be considered:

POOL FIRE: shape
                  diameter
                  drag
                  height

JET FIRE: shape
                  stand off distance
                  length
                  diameter

Note that flame shape will of course be affected by wind: this is discussed in the following section.

It is not the attention to discuss each of these parameters in detail here. However, using pool fire height as an example it is seen from Mudan (1984) that the widely used Thomas correlation bounds 90% of the data presented to within +/-33% accuracy. From Chamberlain (1987) it is seen that a similar degree of accuracy could be expected regarding length prediction for a jet fire model.

It is expected that errors of this magnitude are only likely to be important in the near field. Even so, it should be noted that for a jet fire in particular, such errors may give quite misleading indications regarding the extent to which components are engulfed in flame. This may give rise to quite large errors in predicted times to failure.

3.5 Wind effects

Wind will cause the flame to tilt whether it emanates from a jet or pool fire. For pool fire modelling it seems usual to use the Aga correlation. From the experimental data presented by Mudan (1984) it is deduced that the error on the flame tilt angle if calculated using Aga's correlation is unlikely to be greater than about +/- 20-30%. This would correspond very approximately to an error in flame centre position of about +/- one third of a pool diameter. Such an error could be important but only in the near field. The experimental data of Chamberlain (1987) seems to suggest that a similar sort of scatter is found for jet fires.

Additionally, it is necessary to consider whether or not wind will affect the amount of heat radiated. Cook et al (1987) found no variation in the fraction of heat radiated with wind velocity for natural gas flares. However the same authors note that this conclusion is at variance with earlier published data.

Regarding pool fires, Babrauskas (1983) reports that for 'large' diameter pools (> 1 meter?) burning rate can be doubled by winds of a few metres per second. There appears to be no increase in burning rate as winds increase beyond this velocity range and indeed blow off can occur beyond 5 meters per second. However it should be noted that blow off is likely to be highly dependent on local geometry and pool diameter.

To summarise there seems to be a reasonable understanding of the effects of wind on flame tilt. With regard to burning rate there seems to be greater uncertainty and this is especially so for pool fires. For this latter case it seems that the uncertainty is as high as a factor of two.

3.6 Surface Emissive Power

If utilising a surface emitter model it will be necessary to estimate the surface emissive power (SEP) of the flame. Two approaches are possible here: firstly, if applicable experimental measurements of SEP are available then these can be used. As an alternative, SEP may be estimated by calculating the total heat liberated as radiation and dividing this by flame area (corrections may be included to allow for the fact that SEP will vary over the flame area).

Where direct SEP measurements are used, great care must be taken that correct surface area is considered. Where the second approach is utilised the greatest uncertainty will be associated with estimating the fraction of available combustive energy which is radiated.

The measurements of fraction of heat radiated from jet fires presented by Cook et al (1987) show a scatter of about +/-50% around the line predicting variation with exit velocity. Similar measurements by Chamberlain (1987) show less variation, say +/- 15%.

With regard to pool fires and direct measurements of SEP, the data presented in the SFPE Handbook (1990) suggests that a +/-25% variation in SEP is possible for fires of a given fuel (LNG in this case) and diameter. It seems that, so long as applicable data is available this factor would not be a large source of error.

Much greater uncertainty exists for particularly smoky fires (such as might occur with large diameters and high carbon to hydrogen ratios). The difficulty relates to the amount of heat that is absorbed by the smoke and also the effect of 'radiation bursts' when the smoke clears for a period. Considine (1984) and Mudan (1984) discuss this problem in further detail.

3.7 Radiation Levels Distant from the Flame

Having calculated the radiation emitted from the flame it will usually be necessary to estimate the radiation level incident on a plane at some distance from the flame.

For a surface emitter model the radiation distant from the flame will be estimated using approximate or numerical calculations of the view factor (point source models need only consider square law decay). There is no reason for errors to be introduced at this stage in the analysis.

Uncertainty does arise from the manner in which transmissivity through the air is considered. Transmissivity is principally defined by the extent to which radiation is absorbed by water vapour and to a lesser extent carbon dioxide. The uncertainty is associated with the difficulties inherent in defining the range of wavelengths emitted by the flame and comparing these to the specific bandwidths over which significant absorption takes place. Mudan (1984) proposes a method for resolving these difficulties which assumes that the flame radiates as a black body.

A typical transmission coefficient would be in the range 0.7-0.9 depending largely on flame temperature, distance and humidity. Even if an approximate estimation of transmissivity is made errors should not exceed +/-10%.

3.8 Time-Temperature Behaviour

Given the thermal radiation incident upon a structure it is possible to calculate the time-temperature behaviour of that structure. A number of effects should be taken into account as follows:

- incident radiation

- back radiation

- conduction through the structure and perhaps away to   other structures or surrounding water

- loss of heat by convection

For many problems consideration of the first aspect will be sufficient on its own to permit an adequate understanding of the time-temperature behaviour. Even so all of these aspects are more or less well understood and can be taken into account if required. The conduction problem may, in complicated structural configurations, require the application of a finite element solution. No significant errors should be introduced at this stage if the correct calculation procedure is followed.

The problem is more complex if the flame impinges on the structure. In this case it is necessary to define the heat transfer mechanisms within the flame itself. These will be both radiative and convective. They will vary depending upon their position within the flame. Such variation will be complex an will differ according to the particular situation. For example, there is evidence (see Babrauskas, 1983) to suggest that large pool fires actually burn cooler in their centre due to oxygen starvation effects.

For impingement of a jet fire on a tubular or some such there will be an increase in turbulence which may lead to more efficient burning and an increase in heat release.

Flame impingement effects are, given present knowledge, still a source of great uncertainty.

3.9 Structural Response

Structural collapse may involve failure of process vessels/pipework thereby leading to involvement of further inventory. Alternatively, collapse may involve failure of support structures thereby compromising safe havens, evacuation routes or ultimately platform collapse.

This aspect is rarely dealt with in a satisfactory manner in offshore risk assessments. That this is so is probably due to a combination of factors. The first is probably that the prediction of the behaviour of structures under the imposition of heat loading is a difficult subject. A second factor may well be that the solution of structural problems requires a type of expertise not normally available within safety departments/organisations.

These factors promote the use of the critical temperature approach whereby a member is simply assumed to fail once it reaches a given temperature - typically this might be assumed to be the temperature at which the actual yield stress falls to half of the nominal value at room temperature. Thus failure might be assumed to occur at about 500C.

Unfortunately such an approach may result in times to failure that are significant over estimates. This is because the types of structure found offshore respond in complex ways to the imposition of thermal loading. The principal difficulty arises from the fact that a member cannot be treated in isolation but must be considered as part of the overall structure. As a member is heated it not only weakens due to reduction in yield stress but will also undergo thermal expansion; however such expansion may be resisted by the remainder of the structure thereby increasing the compressive loading in that member. The extent to which expansion is resisted is of course determined by the manner in which loads are redistributed throughout the structure as temperature increases.

At the limit, a fully constrained member may fail at temperatures as low as 140C. Assuming a linear relationship between time and structural temperature (realistic as incident radiation is likely to be the dominant mechanism in the temperature range of interest) it is seen that the critical temperature approach could underpredict time to failure by a factor of four.

Such a large error range is wholly unacceptable. However, relatively sophisticated analysis (see Middleton, 1990) is required to produce a more accurate estimate except in very particular circumstances. The industry awaits a means of simply predicting failure temperature which could be realistically incorporated in a risk assessment.

4.0 CONCLUSIONS

The problems of predicting times to collapse of structures subject to thermal radiation has been reviewed. It is seen that the problem requires a number of analytical steps each of which has an associated error range. The cause and magnitude of the error range varies from step to step. In some instances the potential error arises simply from scatter in experimental data; it is likely that such scatter is inevitable and the associated error in the final result is something that must be accepted.

In other cases the error would arise from an inability to model certain phenomena. The most problematic area in this respect concerns the analysis of structural collapse of frame structures. This topic should be the subject of further research work.

No attempt has been made to estimate the total error in such an analysis (i.e. error bands for calculated time to collapse). This is because not all the steps outlined here are necessarily required for a particular analysis. Furthermore certain errors only apply in particular circumstances (e.g. when considering near field effects). However it is clear that the overall error in such analysis would be non-trivial.

On this basis we must be clear that risk analysis is a tool yielding approximate results which should only be used within an overall decision making framework. This framework should also incorporate engineering judgement and proper consideration of the human element.

REFERENCES

Babrauskas,V., (1983),'Estimating Large Pool Fire Burning Rates', Fire Technology, Vol 19, p251

Chamberlain,G.A. (1987), 'Developments in Design Methods for Predicting Thermal Radiation from Flares',Chem Eng Res Des, Vol. 65, pp299-309

Considine, M. (1984), 'Thermal Radiation Hazard Ranges from Large Hydrocarbon Pool Fires',UKAEA rep.no. SRD R297

Cook,D.K., Fairweather,M., Hammonds,J. and Hughes,D.J., (1987), 'Size and Radiative Characteristics of Natural Gas Flares', Chem Eng Res Des, Vol 65, pp318-325

Crocker, W.P. and Napier, D.H., (1988),'Assessment of Mathematical Models for Fire and Explosion Hazards of Liquefied Petroleum Gases', Jnl of Hazardous Materials, Vol. 20, pp109-135

Knight,F.I., (1983),'Review of the Department of Energy's Offshore Fire Research Programme', Offshore Technology Report OTH 86 229

McMurray,A., (1982),'Flare Radiation Estimated', Hydrocarbon Processing, Nov.

Middleton, C.I., (1990), 'Structural Collapse in Fires - An Overview', conf on Offshore Hazards and Their Prevention, London

Mudan, K.S., (1984), 'Thermal Radiation Hazards from Hydrocarbon Pool Fires', Prog Energy Combust Sci, Vol 10, pp59-80

PHENOMENA	ESTIMATE OF PERCENTAGE ERROR	COMMENT
OUTFLOW RATE, JET FIRES - 1 PHASE	LOW <(+/-10%)?	
OUTFLOW RATE, JET FIRES - 2 PHASE	HIGH >(+/-100%)?	ERRORS CAN BE MADE CONSERVATIVE
BURNING RATE, POOL FIRES	+/-20%	GREATER ERROR FOR SOME CASES - e.g. WAVY WATER
CHOICE OF FLAME MODEL	+/-100%	SURFACE EMITTER MODEL BEST
FLAME SHAPE	+/-30%	ERROR REFERS TO THAT IN FLAME DIMENSION - ERROR IN HEAT RADIATION WILL BE POSITION DEPENDENT
WIND EFFECTS tilt burning rate (pool fires)	+/-25% +/-100%	
SURFACE EMISSIVE POWER	+/-25%	GREATER ERRORS MAY ARISE IN ESTIMATING SURFACE AREA. ALSO IN ESTIMATING SEP FOR VERY SMOKY FIRES
TRANSMISSIVITY	+/-10%?	CAN BE MADE CONSERVATIVE BY ASSUMING TRANSMISSIVITY OF 100%
TIME/TEMP.	+/-10%?	ERROR MAY INCREASE FOR IMPINGEMENT SITUATIONS
BEHAVIOUR STRUCTURAL	UP TO 400%	NEEDN'T BE THIS HIGH BUT ERRORS OF 100% TYPICAL FOR COMPLEX STRUCTURES IF NO DET. ANAL. PERFORMED

TABLE 1: ESTIMATES OF ERRORS ASSOCIATED WITH THE VARIOUS STAGES IN CALCULATING TIME TO COLLAPSE OF STRUCTURES SUBJECT TO THERMAL LOADING

TREATMENT OF ESCALATION MECHANISMS IN THE QUANTITATIVE RISK ASSESSMENT OF OFFSHORE PLATFORMS

by

R.A. Cox and A. Miles
Four Elements Limited
25 Victoria St
London SW1H OEX

ABSTRACT

The Cullen Report has set out a framework for future Safety Cases for offshore installations, in which analyses of fire risks and emergency response, escape and evacuation will play a very important role. The assessment of platform safety and of the need for upgrades to existing installations can only be done in the context of realistic and complete analyses of the potential hazards, and, as the case of Piper Alpha clearly showed, these hazards may manifest themselves through highly complex chains of escalation.

Escalation mechanisms assume much greater importance for offshore installations than they do for shore-based plant, because of the dense packing of hazardous equipment and its juxtaposition with critical control and protection systems. The mechanisms of escalation include: explosion pressure loads, projectile impacts, heat loads from low pressure and high pressure fires, structural collapse, loss of protection systems, power and control.

In this paper, the incorporation of escalation models into offshore platform Quantitative Risk Assessments (QRA) is discussed in general terms. Two approaches are considered: the established Event Tree method, and a novel simulation approach. The importance of the time dimension, both in the physical processes of escalation and in the human activities of emergency response and escape, is emphasised.

1.   **INTRODUCTION**

The Cullen Report has recommended that the future regulation of offshore safety should include provisions requiring for each installation a "Safety Case", which would comprise assessments both of the operator's Safety Management System (SMS) and of the intrinsic hazards and survival capabilities of the platform itself. This paper is only concerned with the latter.

The technical assessment, according to Cullen, would include such things as:

❶   a demonstration that the risks to personnel due to accidental events including hydrocarbon leakage have been reduced as far as is reasonably practicable,

❷   a demonstration by Quantitative Risk Assessment that the risks of damage to the Temporary Safe Refuge (TSR), escapeways, embarkation points and lifeboats are acceptable

❸   a fire risk analysis

❹   an evacuation, escape and rescue analysis.

These are overlapping requirements, as the fire risk analysis would have to be part of any assessment of personnel risks, as would an assessment of evacuation capability. The evacuation, escape and rescue analysis would, in any case, be heavily dependent on the particular accident scenarios considered, which should be generated by prior risk analyses. At present, some operators are therefore treating all of these analyses as an integrated whole, tied together by the framework of QRA, which we consider to be a sensible approach.

It must be remembered that the final legal requirements will be drawn up by HSE, and, based on their experience of the analogous safety regime onshore, they may well adapt, streamline and further define these requirements.

In earlier years, most applications of QRA were for conceptual designs and were used mainly for specifying levels of active and passive protection to be included in the detail design. Now, however, the QRA technique is being widely used by the offshore industry as an aid to decision-making about safety upgrades to existing platforms. Many such upgrades are designed specifically to counteract possible escalation mechanisms. This application therefore calls for an unprecedented degree of completeness, realism and detail in QRA.

For all of the above reasons, it is now very important that escalation mechanisms be properly represented within QRA for offshore platforms. In the following section of this paper, the lessons from Piper Alpha are first considered, while two possible approaches to escalation modelling are treated in Sections 4 and 5.

2. PIPER ALPHA - THE ESCALATION PROCESSES

The case of Piper Alpha illustrates very clearly the variety of escalation mechanisms that may arise during offshore accidents. It is instructive to examine the stages of escalation in this case, as a typical example of the types of phenomenon which a QRA should be capable of representing.

The following stages of escalation probably occurred (Ref: Cullen Report, Chapter 7; approximate times shown at left):

21.58   A relatively small leak of condensate occurred within Module C.

22.00   Condensate vapour or aerosol which had accumulated in Module C ignited at some unknown ignition source, causing a violent explosion.

22.00+   The explosion destroyed the firewall between Modules C and B, and caused major leakage and fire of oil and/or condensate in Module B, probably initially through the failure of the 4 inch condensate injection line, due to impact by firewall debris.

22.00+   The explosion also breached the firewall between Modules C and D, wrecking the control room (direct evidence) and probably knocking out the firepumps, two critical electrical switchboards including the 440v emergency board (indirect evidence).

22.00+   Main power failed, due to a number of possible causes: the main generators could have been damaged directly by the initial explosion; both fuel supplies were lost (the gas because the compressors had tripped earlier, the diesel because the pumps were in heavily damaged areas of Modules C and B); cabling and switchgear was damaged.

22.00+   Emergency power failed. The generator itself was probably able to function, with its local fuel supply intact, but the loss of the 440v emergency switchboard probably tripped it out.

22.00+	The deluge system failed for several overlapping reasons: direct damage to the pumps, failure of main power, pumps switched to manual and no access possible due to heat and smoke, probable damage to ring main and local distribution networks.
22.00+	All primary means of communication failed because of loss of power. Only systems with local battery back-up could operate.
22.01 appr.	Most of the ESD valves closed under fail-safe action, with the probable exception of the MOL valve, ESV 208 (indirect evidence).
22.00 onwards	The wind direction was such that dense smoke from the oil fire in Module B enveloped the living quarters. The LQ exits were blocked by heat and smoke. All lifeboat locations and the helideck were soon cut off by smoke.
22.03 appr.	Oil starts to flow down from Module B to the Cellar Deck (68′ level) through an opening in the floor of Module B where the Main Oil Line penetrates it. This causes an intense localised fire in the vicinity of the Tartan gas riser.
22.20	18" Tartan gas riser fails catastrophically, causing a massive fireball, followed by an intense jet fire.
22.25	First signs of structural failures. East and west cranes collapse, platform decks start to tilt.
22.50	18" gas riser to MCP01 fails catastrophically, causing a third major explosion and further intense fire.
23.00	16" Piper-to-Claymore gas riser fails. Further fireball and subsequent fire.
23.14	Derrick collapsed.

As is well known, the greater part of the structure suffered a progressive collapse over the next few hours, but details of the sequence of the later events are not relevant to this paper. Most of the fatalities arose from smoke ingress into the LQ, where the majority of personnel had sheltered.

There are several lessons from this accident, which are relevant to the physical escalation modelling within QRA. The first is that safety systems may be vulnerable to the effects of relatively small initiating events. The second is that fire may propagate by a great variety of mechanisms: impact by explosion debris; thermal loads on pressurised equipment; gravitational flow of liquids. The third is that smoke movement through modules and around the platform under the influence of the wind can be critical.

Another lesson is the importance of time. Certain escalation processes occurred very rapidly, while others took tens of minutes or even hours to develop. If the wind direction had not been the decisive factor that it was, then the timing of the riser failures certainly would have been, as they would have occurred right in the middle of the evacuation process.

3. The Structure of a QRA for Offshore Installations

Figure 1 shows the principal activities which are carried out in a full QRA. The basic concept has been frequently stated in the past, but this figure seeks to put the emphasis on two particular activities: "scenario generation" and "post-processing". The first of these generates a long file containing descriptions of all possible accident sequences and the corresponding probabilities. The second analyses this file to identify common factors which contribute significantly to the total risk.

For the purposes of this paper, we are primarily concerned with the "scenario generation" stage. There are two basic approaches to this, which are discussed in the following sections, namely the Event Tree approach and the Simulation approach.

4. The Event Tree Approach

In scenario generation, the starting point is a list of "initiating events". These generally fall under the following headings:

o   Releases of hydrocarbons from process equipment
o   Blowouts
o   Releases from risers
o   Ship collisions
o   Structural failures
o   Environmental loads
o   Dropped objects
o   Utilities failures

FIGURE 1 - QRA METHODOLOGY

Platform
description ➡ Identification of
initiating events and
their frequencies

▼ ▼

SCENARIO-DEVELOPMENT
(uses submodels for hazardous
phenomena and behaviour of
platform components)

▼

SCENARIO FILE

▼

POST-PROCESSOR
(analyses scenario file, to
identify dominant contributors
to overall risk)

In practice, the above list of events is expanded into a much longer and more detailed list, specific to each platform. Typically, several hundred initiating events might be defined.

Event trees are developed for each initiating event by drawing a branching line diagram, such as that shown schematically in Figure 2, which details the many possible scenarios that may flow from each initiating event, and their final outcomes. The branch points reflect events which are not deterministic, such as the operation of ESD, deluge activation, and whether or not a release is ignited. By assigning probabilities to each branch of the event tree, the final frequency of each outcome can be established.

Figure 2 is, in fact, a typical example of an event tree used in offshore QRA work. It has seven possible branch points, or "gates", and since each is, in principle, binary, the total number of outcomes is 2^7, or 128. In practice, several of these branches can be ruled out by inspection; for example, "delayed ignition" cannot follow after "early ignition". The number of outcomes may thus be reduced to something more manageable, typically of the order 20-40.

There are several problems with this procedure. The first is the number of gates required to characterise the diversity of escalation pathways that could arise. In the case of Piper Alpha, the relevant gates for the one scenario that actually occurred would have been roughly as follows:

❶	Immediate ignition?	(no)
❷	Delayed ignition?	(yes)
❸	Explosion enough to breach walls?	(yes)
❹	Deluge works?	(no)
❺	ESD works?	(yes - in the main)
❻	Blowdown works?	(maybe)
❼	Wind direction towards LQ?	(yes)
❽	Riser fails?	(yes)
❾	Structure fails?	(yes)

In a full QRA, further gates must be added to these, representing other scenarios that could have occurred (but did not in this case). An example would be local escalation of a fire, without explosion, eventually breaching firewalls or causing a BLEVE.

A second problem with the event tree approach is that the preconception of the sequence of events, necessitated by the act of drawing the tree, excludes certain sequences from the analysis. For example, does the operation of deluge affect the probability and severity of an explosion.....or, does the occurrence of an explosion affect the probability that the deluge will work?

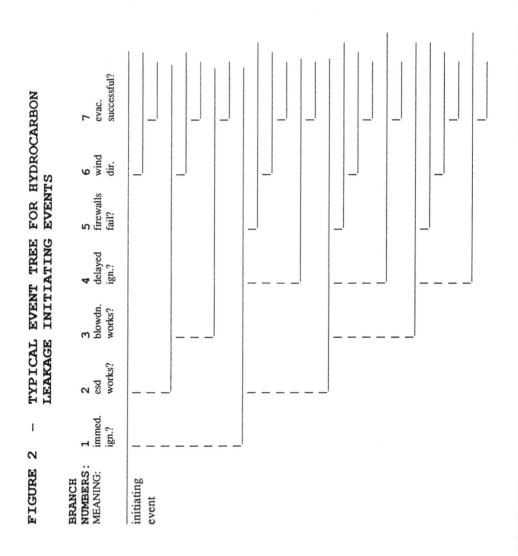

FIGURE 2 – TYPICAL EVENT TREE FOR HYDROCARBON LEAKAGE INITIATING EVENTS

BRANCH
NUMBERS :
MEANING:

1 immed. ign.?
2 esd works?
3 blowdn. works?
4 delayed ign.?
5 firewalls fail?
6 wind dir.
7 evac. successful?

initiating event

A third problem, often encountered by QRA practitioners, is the precise definition of what the gates mean in any one specific case. For example, consider the cases of firewall failure under blast pressure loads, or riser failure under heat loads. It is important exactly which wall, or which riser, has been affected. The event tree fundamentally lacks any geometrical model of the platform - it is simply a probability splitter - the geometrical work must be carried out by the analyst as a side calculation, which will, in fact, constitute the bulk of the work.

Fourthly, there is nothing in the event tree approach that expresses the time dimension - again, that must be added in a side calculation.

Finally, on a purely computational point, the event tree approach is inefficient because it requires the calculation of all of the branches that have been drawn, regardless of their actual significance.

After having spent many years in the practice of QRA for offshore platforms using this approach, the authors have concluded that it has outlived its usefulness as a basis on which to structure a QRA, and have therefore sought a more effective alternative, namely the simulation approach described in the following section of this paper.

5. The Simulation Approach

The basic difference introduced by the simulation approach is that the structure of the event sequences is not pre-conceived, but emerges from the simulated behaviour of each of the platform components.

The particular computational framework that we favour for this problem is "object-oriented" programming, which represents the platform and the accident phenomena as a set of "objects" which interact with each other by passing "messages" in accordance with a pre-determined set of rules. This set of rules is, in fact, identical to the assumptions and sub-models which have to be invoked when doing event tree analysis.

In the present application, the "objects" are:

(i)   Physical components of the platform
(ii)   Events or phenomena, such as a pool fire
(iii)   Groups of personnel
(iv)   The clock

The calculating engine which passes the messages and maintains event logs can be made extremely efficient, robust and general, while all the real modelling is embodied within the object descriptions. This allows the incorporation of sub-models to describe very complex relationships between escalation phenomena and the components of the platform, and the time dimension.

Some of the behaviours will be deterministic; these will result in development of the scenario down a single path. An example of this might be the operation of blowdown removing the inventory from a section of process. Other behaviours may be probabilistic, or at least uncertain in the present state of knowledge; these result in more than one possible state of the platform, which is represented in the simulation by generating a multiplicity of platform state descriptions.

The algorithm that we use automatically selects the branches of highest probability, and the computations are cut off only when it is determined that the aggregate of the remaining scenarios is of negligible significance. There is provision for detecting when a stable state has been reached in any branch of the tree, which terminates further calculation of that pathway.

The time dimension is readily incorporated, by making the object states depend not only upon the messages received from other objects, but also on the time.

The authors have developed a prototype of such a simulation model, which is currently under test. Object models have been developed for the following components, so far:

- Areas of the platform
- Walls
- Vessels
- Compressors
- Valves (as sources of leakage)
- Shut in pipe sections (e.g. risers, manifolds)
- Wellheads
- NRVs, ESDVs, SSSVs, SSIVs, and control valves with an isolation function
- Pipelines
- The reservoir
- Blowdown valves
- Deluge - local area distribution system

FIGURE 3 — EVENT TREE GENERATED BY OBJECT–ORIENTED SIMULATION PROGRAM

Interpretation:

The initiating event is at "0". Subsequent branches are indicated by the numerals. Where there is a numeral, but no branching line, the program has in fact found a branch, but has discontinued its development on the grounds of insignificant probability.

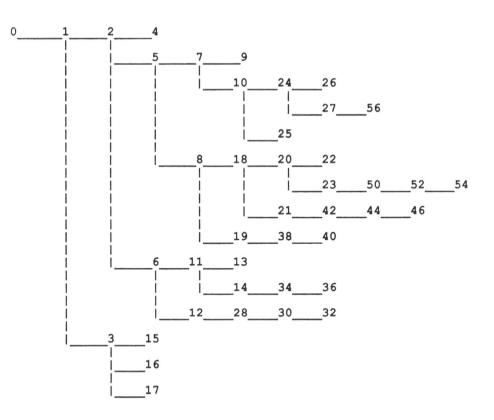

At the time of writing, there is more work to be done before a complete platform can be analysed, however, preliminary simulations have been made on a "platform-like", but incomplete model, and an example of the results is shown in Figure 3, in the form of the equivalent event tree.

The form of this tree is markedly different from what would be seen in any conventional event tree. Many branches have been "pruned" by the frequency cut-off, whilst others have been developed through as many as ten branches, plus a number of deterministic transitions which are not shown on the diagram. The tree also displays one triple branch, which has been generated by combinations of valve closures (the least likely combinations having been pruned).

Another significant difference is that each of the 54 branches is unique. Those that are drawn in vertical alignment are not necessarily identical in meaning, as they would be in a conventional manually-drawn event tree.

This tree synthesis consumed some 2 minutes processing time on a Sun SPARCstation, with full diagnostics on, which indicates that processing time on real problems should be reasonable.

6.   CONCLUSIONS

(i)   For many real applications in the offshore oil industry, such as the assessment of existing installations for such purposes as upgrading, QRA must take proper account of escalation phenomena.

(ii)   Escalation may take a great variety of forms; besides the propagation of fire and structural failure, the survival of active and passive protection systems, and smoke movement, must be taken into account.

(iii)   The event tree method has reached the limits of its usefulness as a concept on which to base the escalation modelling required by QRA; instead, a simulation approach should be adopted.

MODELLING THE BEHAVIOUR OF HYDRO-CARBON LEAKS IN ENCLOSED MODULES ON OFFSHORE OIL PLATFORMS

by

R. Wells, Dr R.H. Jones, Dr C.W. Yip

Technica Ltd

SUMMARY

In quantitative hazard analysis of onshore plant, a release of a hazardous substance is often defined adequately by relatively few parameters. Generally these parameters are the phase of the material, the initial rate of release, atmospheric dispersion factors and the toxic / flammable properties of the substance. Time-dependent behaviour, such as a decaying release rate, is often ignored. The adequacy of such simple release descriptions can be demonstrated by reference to actual accidents that have occurred.

However, if such simple release descriptions are applied in a hazard analysis of an offshore oil / gas installation, then the result of the analysis does not correctly take account of the reliability and effectiveness of such safety systems as gas detection, module ventilation, emergency shutdown and blowdown. All these systems achieve risk reduction because they limit the timescale of the hazard. Therefore, only if the release modelling includes time-varying behaviour can the effect of these key safety systems be assessed.

This paper outlines methods that have been used to model the behaviour of hydro-carbon releases in enclosed, ventilated modules, such as to allow for the successful operation (or otherwise) of the relevant safety systems. It is shown that an adequately realistic representation of the processes can be achieved by a relatively simple set of equations, that can be solved numerically on a personal computer using mathematical modelling techniques.

The results of applications of these models to a number of different installations is summarised. It is found, in general, that the effect of changes to the performance of individual safety systems can be very small, but that the systems act synergistically. For example, there is much more benefit to be gained by improved blowdown if fire or gas detection can be achieved more rapidly. It is also found to be important to avoid the introduction of unnecessary delays to operation of safety systems, in particular if gas explosion risks are to be reduced.

1.0   NOTATION

A	Area of leak (m^2)
C	Gas concentration (fraction)
C_i	Indicated gas concentration (fraction)
C_d	Discharge coefficient of leak path (-)
L	Length of jet flame (m)
LFL	Lower flammability limit of gas (fraction)
M	Molecular weight of gas (-)
N	Number of module air changes (per hour)
P_s	Pressure of process section (Pa)
P(0)	Pressure at time zero (Pa)
P(t)	Pressure at time t (Pa)
Q_f	Mass release rate of fluid (kg/s)
Q_g	Mass release rate of gas (kg/s)
R	Universal gas constant (8314 J/kg/K)
t	Time since start of leak (s)
T	Temperature in process section (K)
V_g	Volume of gas in process section (m^3)
V_m	Volume of module (m^3)
γ	Ratio of gas specific heats (-)
ρ_A	Density of air (kg/m^3)
ρ_g	Density of gas (kg/m^3)
τ	Time constant (s)

2.0   INTRODUCTION

In many applications of risk analysis to chemical and petro-chemical plant, the modelling of the consequences of leaks of hydro-carbons or of other hazardous substances is achieved without any significant consideration of time-dependency. In these analyses, the rate of release of the hazardous material is often calculated simply as a function of the pressure and temperature of the material in store, of the size and shape of the leak path, and of the relevant properties of the material. Thus, for example, the rate of flow of a liquid hydro-carbon is calculated using Bernoulli's formula and the rate of evolution of gas from the liquid is calculated using vapour-liquid equilibrium methods. The fact that the vessel is being emptied by the leak, and therefore that the rate of release will decay, is often ignored.

This simple, time invariant, approach to consequence modelling is very frequently adequate for the purposes of onshore hazard and risk analysis. This is the case because either (i) the entire inventory of hazardous substance is discharged in a very short time-scale, (ii) the rate of decay of release rate etc is negligible or (iii) very serious consequences may occur regardless of the duration of the leak. Examples of actual accidents that can be adequately modelled without considering time dependency are as follows :

> the explosion at the Nypro factory at Flixborough in the UK in 1974. In this case it was concluded at the subsequent Public Inquiry by Parker (1975) that a 20 inch diameter line suffered a full-bore failure. The subsequent rate of release of the flammable substance (cyclo-hexane) was such that the entire contents of the relevant parts of the process were discharged to atmosphere almost immediately.

- the tank farm fire that occurred in 1988 in Singapore. It is found in practise, and Technica (1990) has found by modelling, that the time taken for a single burning oil tank to cause the contents of adjacent tanks to ignite is much less than the timescale of the individual tank fire. Therefore, in modelling such hazards, there is no need to represent any depletion of the inventory of the first tank.

However, safety engineers and risk analysts in the oil industry believe it to be essential to provide measures to limit the duration of accidental releases of hydro-carbons from the production processes on offshore platforms. Such measures include fire and gas detection, emergency shutdown (ESD), emergency blowdown and module ventilation. It appears intuitive that such measures must reduce the fire and explosion risk on offshore platforms.

The conclusion that must be reached from this contradiction between onshore risk analysis and offshore safety engineering is that the time-invariant hazard consequence models applied in many onshore risk analyses are inadequate for offshore applications.

The intention of risk analysis is not simply to provide a measure of the level of risk, to compare against some criterion. It is also the intention of risk analysis to provide a basis against which decisions can be made regarding the optimal configuration of systems, and regarding the choice as to which safety systems should receive the most attention. It is clear that if (for example) gas detection is much more effective at risk reduction than is process blowdown, then relatively more of the available resources should be assigned to the gas detection system. This may be an arbitrary example, but it is this sort of decision process that the results of risk analyses should be able to assist. Such decisions can clearly only be made using risk analysis if the results are based on modelling that takes correct account of the effect of the systems being considered.

From the above it may be concluded that it is essential that the hazard consequence modelling that is applied to hydro-carbon releases in offshore platform modules must allow for at least all the following effects :

- the rate of accumulation of gases, under the influence of forced ventilation,

- the speed of response of gas detectors to the rise in gas concentration, and the speed of response of flame detectors to ignition of the hydro-carbons,

- the speed of closure of ESD valves and the speed of opening of blowdown valves,

- the rate of decay of system pressure, under the influence of the leak and the influence of blowdown,

- the rate at which the accumulated gases disperse due to module ventilation, and

- the effect of changes that may be made to the module ventilation rate, on detection of the released gases.

In section 3, a methodology is proposed that in principle can account for all of the above factors. In section 4, examples are given of application of these modelling methods to a range of offshore installations.

3.0 METHODOLOGY

3.1 Scoping Calculations

Explosion Hazard. It is known from experience, and Bakke (1989a) has shown by scale model tests and by computer simulation, that the ignition of a module full of gas between the flammability limits can have very severe consequences. Bakke (1989b) has also shown by model tests that the accumulation of gases, when released at high pressure into a module, is effectively uniform. Therefore, the accumulation of gases throughout the module, at their lower flammability limit (LFL), is taken to represent the hazard for the purposes of these scoping calculations.

Typical offshore platform enclosed modules have volumes between 2000 and 5000 cubic metres and a forced ventilation system that achieves a rate of air change of about 12 per hour. Typical released gases will have LFL values of between 0.02 and 0.05 by volume, with these values corresponding to average gas molecular weights of 50 to 20 respectively.

The mass release rate required to overcome the diluting effects of module ventilation system, such that a widespread flammable concentration of gas can accumulate, can be calculated as

$$Q_s = \frac{LFL . N . V_m . \rho_g}{(1 - LFL) . 3600} \quad kg/s \tag{1}$$

For the above data, the value of this limiting (critical) release rate is found to be between about 0.3 and 1.0 kg/s. For normal ventilation, the actual value for the limiting release rate, for any one case, is found to be most strongly influenced by the module volume.

At flow rates just above these limiting values it will take a significant time for the gas to accumulate, which would allow time for the shutdown and blowdown systems to act to prevent such an occurrence. However, it does not require leak rates much in excess of these limiting values to produce a widespread flammable gas accumulation over a much shorter time-scale, such that the speed of response of the protection systems becomes critical. At still higher release rates, the protection systems cannot respond in time.

Fire Risk. The length of a jet fire of typical hydro-carbons, released at high momentum, may be calculated approximately by

$$L = 15 . Q_f^{0.4} \quad metres \tag{2}$$

This is a modified version of the well-known correlation derived by Wertenbach (1971), that has been found to apply to a wider range of release types, including both gas and two-phase mixtures. From this correlation, it can be seen that a release of hydro-carbon at a rate of only about 1 kg/s will, if ignited, produce a flame of length about 15 metres. In any typical offshore module, it is highly likely that such a flame will impinge on some critical plant or structural item. Cowley (1990) has shown that the rate of transfer of heat from a turbulent jet flame to any item on which it impinges can be of the order of 250 kW/m^2. At such rates of heat transfer it takes only a short time for significant damage to result. Therefore it is critical for minimisation of the fire hazard that rates of release of greater than, or around, 1 kg/s must not be sustainable for any significant period.

Conclusion. In typical large offshore platform modules, leaks at a rate of substantially less than 1 kg/s (say 0.1 kg/s) are unlikely to give rise to significant explosion hazards. In addition, the fire hazard from such small leaks should be controllable by fire-fighting systems. Leaks of around 1 kg/s may have serious potential fire and explosion consequences, and these hazards therefore must be controllable by the action of the protection systems. Leaks of a much greater rate than 1 kg/s will almost inevitably give a serious hazard potential, regardless of the action of protection systems. The three ranges of leak rate, and their possible resultant consequences, are shown graphically in Figure 1.

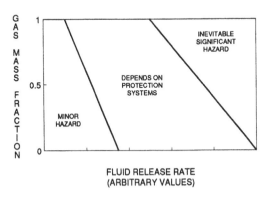

Figure 1 : Leak Rate Categorisation

The effect of improved module safety system capability (e.g. higher ventilation rates, faster blowdown) is to move the category boundaries to the right in the Figure. The effect of improved reliability of safety systems is to reduce the likelihood that "controllable" hazards actually become out of control.

Given that the safety systems are expected to have their greatest effect on leaks in the region of 1 kg/s, it is important that the modelling of leak consequences is appropriate for such rates of release.

3.2   The Model Equations

The model comprises a set of simultaneous equations, representing each of the various phenomena involved in the leak consequences. The model equation set is as outlined below.

Fluid Release. The rate of release of fluid is calculated using one of the standard formulae. For gas leaks, the formula is as follows :

$$Q_g = C_d A P_s \left[\frac{\gamma M}{RT} \left(\frac{2}{\gamma + 1} \right)^{\frac{\gamma + 1}{\gamma - 1}} \right]^{\frac{1}{2}} \tag{3}$$

Bernoulli's equation is used for leaks of liquid, and one of the various alternative correlations is used for leaks of two-phase materials.

System Pressure. Up until the instant the process is shut-down, the system pressure is assumed to remain constant. This assumption may be seen to be valid in the range of interest, as the rates of release that are being considered are much less than the typical throughput of an offshore process system. Therefore process pressures etc will not be much affected by the leak until system ESD occurs. Once the process is shut-down, the system pressure is assumed to decay. For systems containing only gas the rate of decay is :

$$\frac{dP_s}{dt} - \frac{-RT}{MV_s} Q_g \quad Pa/s \tag{4}$$

The model assumes that the system is isothermal and that gas properties do not change with changing pressure. Equation (4) gives an exponential decay in pressure, of the form $P(t) = P(o) \exp(-t/\tau)$.

For systems containing both gas and oil, the pressure decay is more difficult to model, and the form of the decay depends on whether the leak is from the part of the system that contains the gas or from the part of the system that contains the liquid. However, in no case does it prove to be impossible to model the decay by a relatively simple set of equations. The least tractable case is one where the system contains only a volatile oil. In this case, depressurisation down to the bubble point occurs very rapidly, due to the low compressibility of liquids. Subsequent to reaching the bubble point, depressurisation is slower and during this stage a homogeneous equilibrium state is assumed. The sudden depressurisation, and the sudden change to a different model form, requires special care in the numerical solution of the equation set.

The volume term in this part of the equation set is the volume of the section of the process that is leaking, thereby taking account of the fact that ESD actions will have isolated this part of the process from all others. The effect of failure of isolation can be modelled by assuming a larger value for the system volume.

Gas Accumulation. This part of the model set is only used if it is assumed that ignition of the release does not occur before the gas detection system signals ESD.

The rate of change of gas concentration in the module atmosphere is given by

$$\frac{dC}{dt} - \frac{Q_g - C(Q_g + NV_m \rho_A/3600)}{\rho_A V_m} \tag{5}$$

The above model is very similar to that of the US National Fire Prevention Association (1981). However, in this case a higher degree of mixing is assumed which, as stated above, has been verified by experiment.

In the case of releases of liquid, which subsequently flash off some gas, the flow rate term in the above equation is the flow rate for the evolved gas only.

Flame Modelling. For gas releases, and for liquid releases that have high flash, it is generally assumed that a jet fire will result and this is modelled using equation (2). For releases of low volatility liquid, a pool fire is more likely and this is modelled by (for example) the models given by Moorhouse (1982). The rate of heat release from the flame can be calculated easily for both cases, again using standard correlations.

Modelling ESD. If ignition of the hydro-carbon occurs before the high gas levels are detected, then it is assumed that ESD is triggered immediately. However, triggering of ESD by gas detection is more difficult to represent. This is described below.

Manufacturers data on the sensitivity of catalytic-type flammable gas detectors, which correspond closely to the theoretical values given by Firth (1973), are as follows :

Hydrocarbon	Methane	Ethane	Propane	Butane	Pentane
Relative Sensitivity	1.0	0.5	0.45	0.40	0.35

The response of detectors to a given concentration of gas is seen to depend on the composition of the gas. As can be seen from the table, the relative response is lower for the higher hydro-carbons. Typically, the detector will be calibrated using methane, and the upper (ESD) alarm level will be set at between 30% and 60% of the methane LFL. However, the ESD level is then between 60% and 120% of the ethane LFL, and is between 90% and 180% of the pentane LFL, for example. The benefits of low alarm settings are apparent, especially for cases where the released gases may contain large fractions of higher hydro-carbons.

Furthermore, the detectors have a finite response time. The basic detector (pellister and housing, including the sinter element) would typically take about 10 seconds to give a reading of 90% of the final value when subject to a step increase in gas composition. However, with the addition of weather protection this response time can increase to 60 seconds or more, depending on the rate of air movement past the detector.

The first factor is modelled by the following:

$$\frac{dC_i}{dt} = \frac{C - C_i}{\tau} \tag{6}$$

ESD is then assumed to be triggered when the indicated gas concentration, C_i, reaches the pre-set alarm level, modified to allow for the detector sensitivity.

In general, it is assumed in the modelling that ESD valves take a finite time to begin to close after the triggering of ESD, but that the closure then occurs very rapidly.

Blowdown. In general, blowdown valves are only opened after some pre-set delay following ESD. Such delays can be allowed for in the modelling. Once the blowdown valves are open, their effect is modelled by adding another term to the leak rate in the equation for system pressure.

Solution of the Equation Set. It has been found that models such as that outlined above may be solved on a typical personal computer by any of three methods. These are : purpose-written Fortran computer programs, general-purpose simulation languages, and spreadsheet packages such as Lotus 1-2-3. It has been found that each of these methods has advantages and disadvantages, but that each of them is capable of giving accurate results in a few seconds computation time, with a high degree of flexibility. Use of spreadsheet programs, or of general-purpose simulation packages, is found to be more efficient than writing of special purpose code.

34

4.0 TYPICAL RESULTS

4.1 Gas Compression System

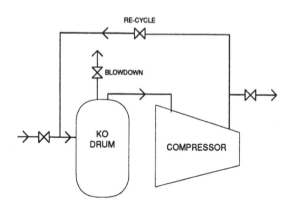

Figure 2 : Simplified Schematic of a Gas Compression System

The overall system, as it is considered here, comprises the compressor itself, a suction knock-out drum, ESD (isolating) valves on the inlet to the suction drum and on the compressor discharge, and a blowdown valve on the suction drum. This case is made more complicated, as there is another action taken on fire or gas detection, in addition to ESD and blowdown. This is that a re-cycle line is opened between the inlet and outlet of the compressor. Once this valve is opened, there will be flow between the compressor discharge and suction sides. Normally, this flow will be from the discharge to the suction. However, it is possible for this flow to reverse. Typical results of simulation of compression system leaks are given below :

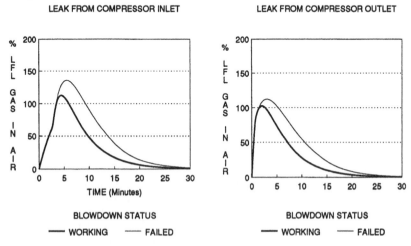

Figure 3 : Simulation Results for Gas Compressor Leaks

Values of key parameters in this hypothetical example are:

Inlet pressure	20 bara	Outlet pressure	150 bara
Inlet volume	5 m³	Outlet volume	2.5 m³
Module volume	3000 m³	Air change rate	12 / hour
Gas LFL	0.05	ESD level	0.03
Detector Lag	5s	Blowdown delay	30s
Leak diameter	20mm	Blowdown area	500 mm²

In this example it is found that blowdown failure is not highly critical. However, blowdown can be very important, depending on the values of key system parameters. It is of note that the hazard from the (lower pressure) suction leak is at least as great as that from the (higher pressure) discharge leak, due to the action of the re-cycle valve.

4.2   Separation System

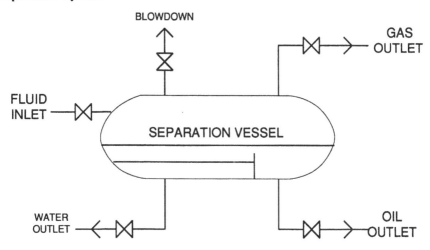

Figure 4 : Simplified Schematic of a Separation Vessel

For the purpose of this modelling, the separator is considered to comprise simply a vessel, approximately half filled with liquid, padded by evolved gas and having isolatable tappings for fluid inlet, gas outlet and liquid outlet. There are two different cases of interest for a separation system leak. These are leaks from the vessel gas space and leaks from the vessel liquid space. Results obtained for a gas leak from a typical separation vessel are as shown in Figure 5.

Values of key parameters in this example are:

System pressure	7 bara	Gas flash fraction	5%
System volume	70 m³	Liquid fill	50%
Module volume	2750 m³	Air change rate	12 / hour
Gas LFL	0.03	ESD level	0.025
Detector lag	5s	Blowdown delay	30s
Leak diameter	35mm	Blowdown area	1000 mm²

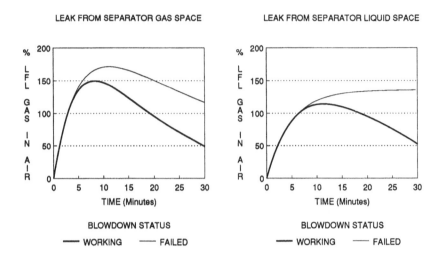

Figure 5 : Simulation Results for Separation System Leaks

This example shows how the operation of blowdown is often relatively ineffective in mitigating separation system leaks. This results from the fact that the liquid in the vessel will flash off large quantities of gas as the pressure in the vessel falls, and this gas has also to be discharged from the vessel, either by the leak or by blowdown.

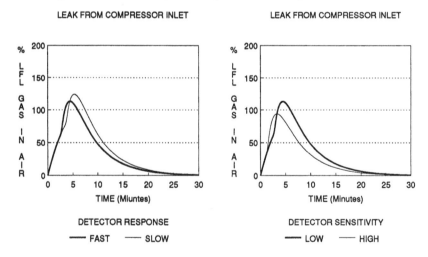

Figure 6 : Gas Accumulations for Different Detector Responses

The above figure shows the effect of changes in the sensitivity and response times of gas detectors. The model is basically the same as used to produce Figure 3. In each graph the base case is one where it was assumed that ESD is initiated at 2.5% gas in air, and that the gas detectors have a fast response time (5 seconds lag). The other cases show the changed gas accumulation that can result from (i) a lower alarm set-point (1.25% gas in air) or (ii) a slower gas detector response (60 seconds lag), such as would occur with the use of weather shielding.

4.4　The Effect of Improved Ventilation

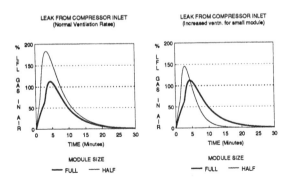

Figure 7 : Gas Accumulations for Different Ventilation Cases

The above figure shows the effect of changes in module ventilation rate. The model is basically the same as used to produce Figure 3. In the left-hand figure, the moduel volume is halved (to 1500 m³) without any increase in rate of air change per hour. In the right-hand figure, the module volume is halved but the rate of air change per hour is doubled (i.e. the same volumetric air flow rate is achieved). The significance of these cases arises from the fact that the rate of accumulation of gas in a module does not depend solely on the rate of air change, but also depends on module volume. This can be seen from equation (5). However, this factor is often ignored, and the adequacy of ventilation is usually determined by achievement only of a standard number of air changes per hour. Given a defined leak scenario, a much greater level of gas accumulation results if the module is of a smaller volume. To restore the level of hazard, the module ventilation rate (in terms of air changes per hour) has to be increased accordingly. As has been concluded by Gale (1985), ventilation rates should ideally be specified in terms of a volumetric rate of change (m³/s), rather than as a number of changes per hour.

5.0　APPLICATION TO NATURALLY VENTILATED MODULES

In principle, there is no reason why the above modelling methods cannot be applied to open, naturally ventilated, offshore models. Indeed, the authors have done so. The main difficulties that arise in such applications are (i) estimating the rate of ventilation of such modules, as a function of wind speed and direction, (ii) carrying out sufficient simulations that a realistic cross-section of ventilation rates are covered and (iii) deciding whether the ventilation is even over the entire module or whether it is necessary to consider that the gas accumulates in a portion of the module only.

These factors can be critical. For example, in the report of the Inquiry into the Piper Alpha Disaster, Cullen (1990) concludes that the initial accumulation of gas probably occupied less than about 25% of the relevant module volume. In an Appendix to his report Cullen (1990) also describes wind-tunnel tests that showed an air change rate at the time of the disaster of 39 per hour, in a wind speed of 8.2 m/s (15 knots). This shows a high rate of air change, compared to that from mechanical systems, even in relatively calm conditions. However, the reduced mixing volume results in more rapid accumulation of a large-scale flammable mixture of gas in the module than would occur otherwise.

6.0   CONCLUSIONS

The modelling of the consequences of hydro-carbon leaks in offshore modules must take account of the time-dependency of the leak and of the other important factors. The rate at which the leak consequences decay is critical to the degree of hazard that the leak presents.

If hazard and risk analysis is to be of assistance to the specification of fire / gas detection systems, of ventilation systems, and of ESD and blowdown systems, then the effect of these systems on leak consequences must be included in the consequence modelling. This strengthens the need for the modelling to include time dependency.

The complexity of the consequence modelling does increase as a result of the above. However, it has been found that the equation sets are generally sufficiently simple that they may be solved on a modern PC within a few seconds, using either purpose-written software, general-purpose simulation packages or spreadsheets. All the results presented in the main body of this paper have been generated using a PC-based general-purpose simulation language. An example of the output from these models, when coded onto a modern spreadsheet package, is shown in Appendix I.

It has been found that the risk benefit of fire / gas detection etc is best explained by considering all leaks to fall into three size categories. The leaks in the smallest size category cannot produce a significant degree of hazard, because the ventilation system will prevent any large accumulations of hazardous gas and because any resultant fire is likely to be controllable. These leaks have a rate of release that is typically much less than 1 kg/s. The leaks in the largest size category will always produce severe consequences (if ignited), regardless of whether safety systems operate. This is because the ventilation system cannot prevent significant gas accumulation, and because any resulting fire will be massive. Whether the leaks in the middle category produce severe consequences depends on whether the safety systems operate. These leaks have a rate of gas release that is of the order of 1 kg/s. The capacity of the safety systems determines the boundaries between the three leak categories. The reliability of the safety systems determines the outcome of leaks in the middle size category.

A rate of gas release of the order of 1 kg/s might result, depending on system pressure etc, from leaks of a diameter of 10mm or less. The critical leak sizes for the examples presented in this piper are larger than this, but this simply reflects the relatively low system pressures assumed in the example simulations. The critical nature of 10mm diameter or smaller leaks is significant as this corresponds, for example, to a sheared instrument connection or to a relatively minor flange failure. Such leaks clearly are very credible.

In the majority of the studies carried out using the methodology described in this paper, it has been found that the safety systems are synergistic. That is, the effectiveness of one safety system is very strongly influenced by the effectiveness of other safety systems. For example, it is found that there is little benefit in having rapid acting ESD valves and a high capacity blowdown system if the gas detection system has a slow response or a low sensitivity.

The most important individual results found during application of these methods, are as follows:

- the required capacity of ventilation systems for enclosed modules is better based on a volumetric air change rate (i.e. in terms of cubic metres per second) than on a number of overall module changes per hour. Otherwise, very small modules may be provided with ventilation systems that are incapable of diluting to safe levels any but the smallest of hydrocarbon gas releases.

- hydro-carbon gas detectors should be set to give high alarms at no more than about 30% of the LFL of methane, and at lower levels if at all possible. Otherwise they provide an inadequate response to heavier hydro-carbons and are much less likely to be able to signal ESD sufficiently promptly to prevent hazardous accumulations of flammable gas.

all gas detection, ESD and blowdown systems should be arranged to respond as rapidly as possible to the release of hydro-carbons. Potential causes of degraded response of these safety systems are the use of weather shielding on gas detectors and the incorporation of unnecessary delays between ESD and blowdown.

7.0 REFERENCES

Bakke, J.R., (1989a) 'Practical Applications of Advanced Gas Explosion Research, FLACS - A Predictive Tool', 6th International Symposium on Loss Prevention and Safety Promotion in the Process Industries, European Federation of Chemical Engineers, Oslo, Norway.

Bakke, J.R., (1989b), in the Piper Alpha Inquiry Transcripts, Day 77, Page 29, Aberdeen, UK.

Cowley, L.T. and Pritchard, M.J., (1990), 'Large-Scale Natural Gas and LPG Jet Fires and Thermal Impact on Structures', GASTECH 90, Amsterdam, Netherlands.

Cullen, The Hon Lord, (1990), 'The Public Inquiry into the Piper Alpha Disaster', Department of Energy, HMSO.

Firth, J.G. et al, (1973), 'The Principles of the Detection of Flammable Atmospheres by Catalytic Devices', Combustion and Flame, Vol 21.

Gale, W.E., (1985), 'Module Ventilation Rates Quantified', Oil and Gas Journal, Dec 1985.

Moorhouse, J., (1982), 'Scaling Criteria for Pool Fires Derived from Large Scale Experiments', I. Chem. E. Symposium Series no. 71, 165-175.

Parker, R.J., (1975), 'The Flixborough Disaster. Report of the Court of Inquiry', HMSO, London.

Technica, (1990), 'Atmospheric Storage Tank Study', Report to Oil and Petrochemical Industries Technical and Safety Committee (Singapore).

TNO, (1976), 'Methods for the Estimation of the Consequences of the Physical Effects of the Escape of Dangerous Materials (Liquids and Gases)', Voorburg, Netherlands.

US National Fire Prevention Association, (1981), 'Explosion Prevention Systems', in US National Fire Code no 69-1, Boston, USA.

Wertenbach, H.G., (1971), 'Spread of Flames on Cylindrical Tanks for Hydrocarbon Fires'. Gas und Erdgas, 112(8), 383.

41

APPENDIX I : EXAMPLE OUTPUT FROM SPREADSHEET MODEL

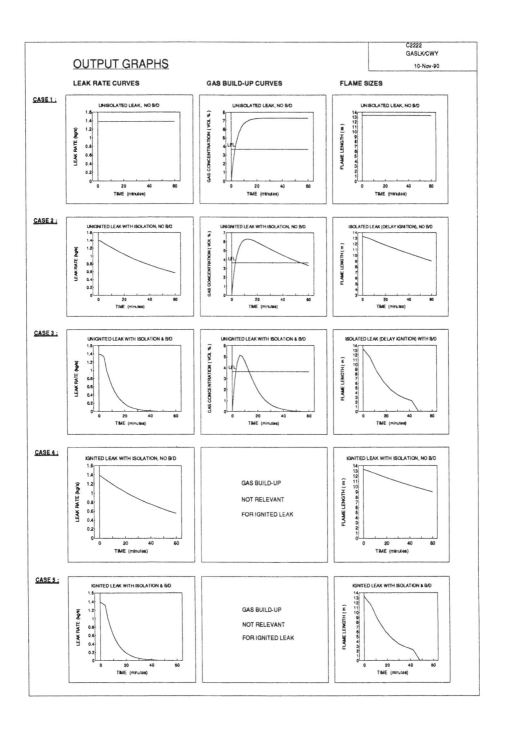

AN INTEGRATED APPROACH TO FAULT TREE ANALYSIS FOR SAFETY AND AVAILABILITY STUDIES

D J Burns (WS Atkins Engineering Sciences Limited, UK)

SUMMARY

The application of Fault Tree Analysis (FTA) to offshore and onshore installations is a key step in identifying inherent weaknesses in design or procedures which could have serious consequences for the safety or availability of the installation. The roles of the operator, the equipment vendor and the reliability engineer are discussed with a view to setting and achieving targets for safety and availability.

A means of assessing the performance of the installation against set targets of safety and availability is described, using an integrated package for fault tree construction, cut set analysis and post-processing.

INTRODUCTION

Among the numerous criteria which must be met by hydrocarbon processing or handling facilities, two are discussed in this paper where the use of Fault Tree Analysis (FTA) can be of great assistance (Watson 1989).

The first criterion is that of safety and, specifically, the need to demonstrate that hazardous events can be safely contained by reliable contingency operations and systems. The risk of damaging consequences to human beings, plant or environment must be shown to be acceptably low. Thus Section 2 presents a schematic model for the main steps employed in a Quantitative Risk Assessment (QRA), indicating where FTA is applied.

The second criterion is that of plant operational availability, and specifically, the need to demonstrate that the availability targets can be met at, or near to, the optimal cost for the life time of the plant. Section 3, therefore, presents a further schematic model where the emphasis is on availability of plant and sub-systems, and on the interests of both operator and equipment vendors in demonstrating that the availability criteria will be met.

The role of availability in the Life Cycle Cost (LCC) of a plant is discussed in Section 4.

Section 5 comprises an overview of an integrated suite of programs for FTA and LCC analyses which has been developed with the above needs for safety and protection of the investment in mind.

QUANTITATIVE RISK ASSESSMENT

The essence of this type of study is to demonstrate that a plant is safe by assessing the level of risk associated with all identifiable major hazard events. The risk is generally stated as an estimated frequency of occurrence for a certain level of damage. The level of damage lies within the domain of consequence analysis, and will not be addressed here. However, the estimation of frequency of occurrence is one of the uses of FTA (Hirschberg S and Knochenhauer M 1989, Bjore S et al 1988, Hirschberg S et al 1988).

Figure 1 shows the main building blocks of a QRA. After ascertaining the workings of the plant, the accident initiating events are identified by various techniques such as Hazard and Operability Study (HAZOP), Failure Modes Effects and Criticality Analysis (FMECA) and surveys of case histories.

For each identified initiating event, an event tree is constructed whereby the worst possible accident scenarios are postulated. At each branch in the event tree, an event is defined which can aggravate the scenario if it occurs. Accident scenarios leading to Major Catastrophes are said to be initiated by Major Hazard Events. These are then quantified on two counts: firstly that their damage effect is calculated by physical models, and secondly that their frequency is estimated. This is often achieved by FTA, when the event is broken down into possible precursors, the estimated frequencies of which are combined using Boolean logic.

It has been noted that the event tree contains postulated aggravating events, whose probability of occurrence needs to be calculated in order to arrive at a final frequency estimation for the catastrophic event. Again FTA is an ideal means of arriving at branch probabilities.

The event tree analysis is carried out for several initiating events, and the frequencies of all like catastrophic events are summed from all initiating event considered in order to make comparison with acceptance criteria. Plants which do not meet the criteria will need to have some redesign, if at the design stage, or, if operational, some back-fitting.

PLANT AVAILABILITY ASSESSMENT

Availability analyses of plant are often carried out as a function of time by simulation techniques. However, mean unavailability over a

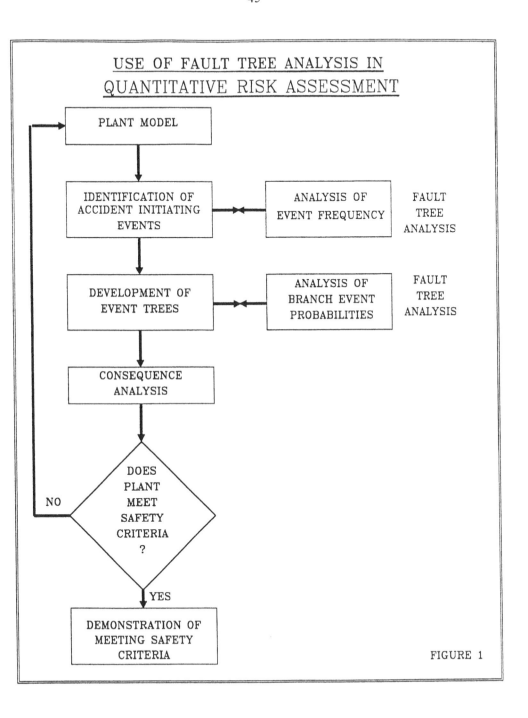

USE OF FAULT TREE ANALYSIS IN QUANTITATIVE RISK ASSESSMENT

FIGURE 1

period of time can be estimated using FTA, and this is useful for vendors wishing to demonstrate the total availability of their systems or to optimize redundancy in equipment or spares holding (Hirschberg S et al 1988, Knochenhauer M et al 1989).

Figure 2 shows the scheme for applying FTA to availability modelling. Starting with the plant model, more than one operational mode may be possible, the availability target for each operational mode being different. For each operational mode, a fault tree top event will be definable reflecting the frequency of failure of the plant. A fault tree can then be drawn up to indicate the possible causes of total plant failure which, when provided with failure and repair data for all basic system failure events will constitute the Integrated Plant Unavailability Model.

Each system failure is then made the top event of a separate fault tree, and a break-down of system failures into component failures carried out. As availability targets can be determined for each system, in order to meet the total plant availability target, it is possible to present vendors with availability targets for their equipment. In some cases the initial target established by the operator cannot be met by the vendor without extra cost, and negotiations may result in a compromise being reached. The FTA is very useful here in demonstrating the sensitivity of the total plant availability to each system's performance. So not meeting the original system availability target set by the operator could result in

a)   resetting the target for the system availability
b)   redesigning the system to meet the original target
c)   redesigning the plant to meet the availability target

Resetting the plant availability target is possible, but unlikely. The above procedure would be repeated for each operational mode.

AVAILABILITY AND LIFE CYCLE COST (LCC)

In additional to demonstration of a particular vendor's system availability, the operator will wish to calculate the total cost of procuring equipment, running and maintaining the plant, and of production loss when the plant stands idle (Hirschberg S et al 1988, Knochenhauer M et al 1989).

System designers, reliability engineers and procurement staff should work together to arrive at the cost-optimized availability goal, taking into account initial equipment costs, levels of redundancy, maintenance costs, and cash flow.

The relationship between the parameters involved in LCC considerations is shown in Figure 3.

Availability of operation can, in theory, be increased more and more by investing in more and better equipment. Conversely, at a low level of investment cost, more operational costs are incurred due to plant breaking down. As investment increases, so the need for maintenance (operation costs) decreases. Thus the total cost (LCC) passes through a minimum. The reliability engineer, as coordinator between design and procurement, can assist greatly in getting the availability target near to the minimum LCC.

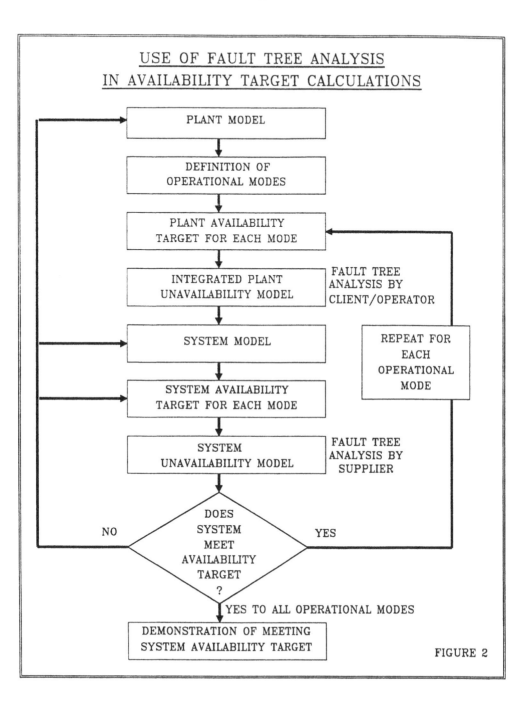

USE OF FAULT TREE ANALYSIS
IN AVAILABILITY TARGET CALCULATIONS

PLANT MODEL

DEFINITION OF
OPERATIONAL MODES

PLANT AVAILABILITY
TARGET FOR EACH MODE

INTEGRATED PLANT
UNAVAILABILITY MODEL

FAULT TREE
ANALYSIS BY
CLIENT/OPERATOR

SYSTEM MODEL

REPEAT FOR
EACH
OPERATIONAL
MODE

SYSTEM AVAILABILITY
TARGET FOR EACH MODE

SYSTEM
UNAVAILABILITY MODEL

FAULT TREE
ANALYSIS BY
SUPPLIER

DOES
SYSTEM
MEET
AVAILABILITY
TARGET
?

NO

YES

YES TO ALL OPERATIONAL MODES

DEMONSTRATION OF MEETING
SYSTEM AVAILABILITY TARGET

FIGURE 2

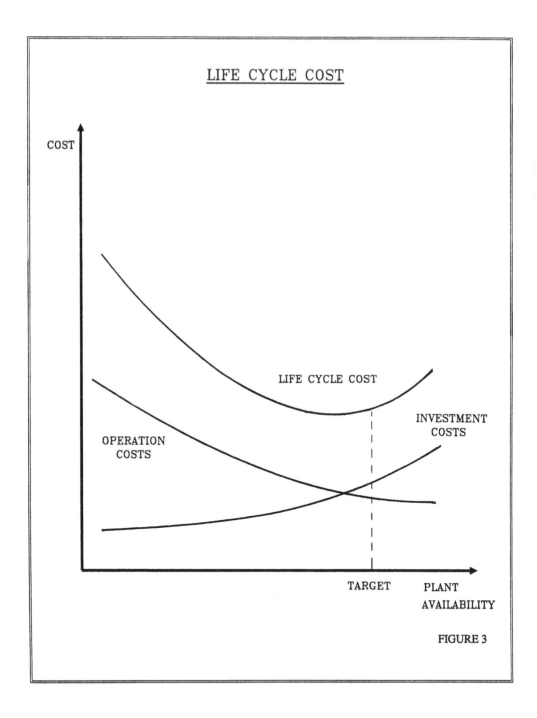

FIGURE 3

COMPUTERISED FAULT TREE AND LCC MODELLING)

The assessment of availability and its application to targets of safety, plant operability and LCC may be carried out quickly and effectively using the SUPER-NET package. Developed by ABB Atom in Sweden, this consists of the following units (Figure 4):

SUPER-TREE	for screen-orientated fault tree handling
CUTSET	for fault tree analysis
SENS	for importance and sensitivity analysis
FRANTIC	for time-dependent reliability analysis
SAMPLE	for statistical uncertainly analysis
COST	for Life Cycle Cost analysis

SUPER-TREE

This is a semi-automatic fault tree handling program which allows the fault tree to be built up interactively on the screen of a PC or Minicomputer. The tree structure is left-adjusted to enable an automatic assignment of gate addresses (Figure 5). Whole sections of the tree can be copied and relabelled automatically and checks are in place for errors in the tree structure. Drawing and restructuring of the tree can be carried out at two levels of detail, while a third level is reserved for details of basic event data including failure rate, repair time and cost and replacement cost. The data can be transferred automatically to the basic event in the fault tree by an event code system, or manually as required. The various levels are illustrated in Figure 6.

CUTSET

Top events of fault trees represent either the frequency of some event, such as total plant failure, or the failure on demand of a system or piece of equipment. Whichever type is under analysis, the combination of events which bring about the top event are called cut-sets, and the numerical analysis of the, often many, combinations of events may be carried out using a Boolean reduction by the CUTSET program. The cut-sets are presented in order of magnitude, and are easily identified by the user-specified event coding system. The total unavailability of frequency of failure for the top event is presented as the sum of the individual cut-set values.

SENS

The cut-sets, as calculated according to the preceding section, are summed to give a first-moment estimation of the total unavailability or failure frequency. This approach assumes no dependence between the cut-sets and is therefore an approximation, which is satisfactory providing no significant level of interaction between the failure modes, such as common causes, is applicable. If such an interaction is considered to be valid, more precise results may be generated using the SENS program. This performs sensitivity analysis and lists importance rankings on the results from the CUTSET analysis.

For the base case, Fussel-Veseley importance measures are generated for all the basic events in the cut-set list. This is then repeated by

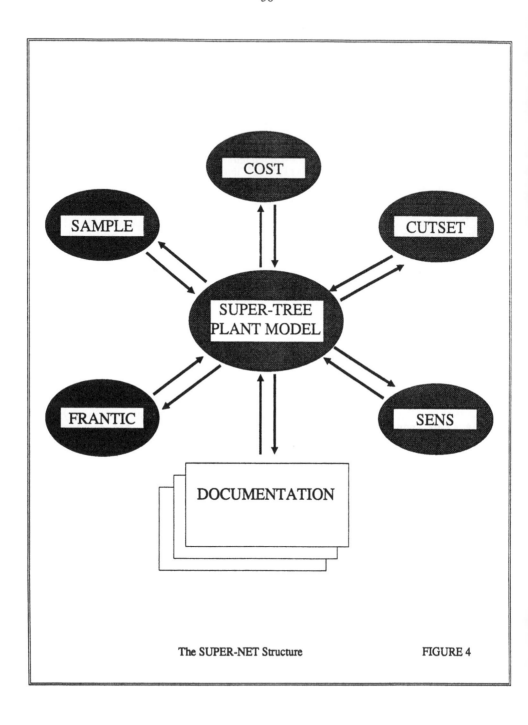

The SUPER-NET Structure   FIGURE 4

51

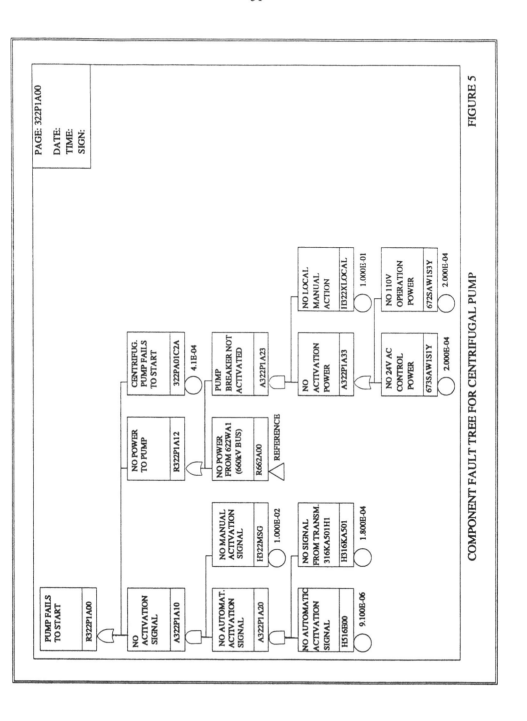

COMPONENT FAULT TREE FOR CENTRIFUGAL PUMP

FIGURE 5

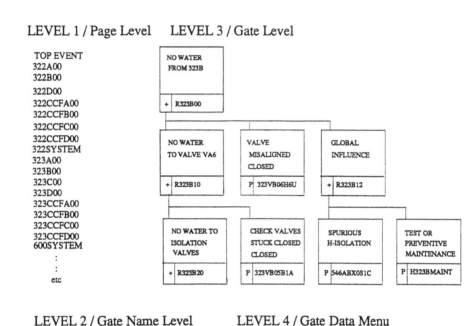

LEVEL 1 / Page Level

TOP EVENT
322A00
322B00

322D00
322CCFA00
322CCFB00
322CCFC00
322CCFD00
322SYSTEM
323A00
323B00
323C00
323D00
323CCFA00
323CCFB00
323CCFC00
323CCFD00
600SYSTEM
:
:
etc

LEVEL 3 / Gate Level

NO WATER FROM 323B
+ R323B00

NO WATER TO VALVE VA6	VALVE MISALIGNED CLOSED	GLOBAL INFLUENCE
+ R323B10	P 323VB06H6U	+ R323B12

NO WATER TO ISOLATION VALVES	CHECK VALVES STUCK CLOSED CLOSED	SPURIOUS H-ISOLATION	TEST OR PREVENTIVE MAINTENANCE
+ R323B20	P 323VB05B1A	P 546ABX0S1C	P H323BMAINT

LEVEL 2 / Gate Name Level

R323B00

R323B10   323VB06H6U   R323B12

R323B20   323VB05B1A

R323B30   323VB04M3A

R323B40   323VB03B1A

R323B50   R323P1A00

R323B60   323VB01M4D

316C000T1Y   323CX01S1X

546ABX0S1C   H323BMAINT

LEVEL 4 / Gate Data Menu

Gate name	:323PA01C2A0302
Diamond flag	:N
Diamond mode name	:
Atribute line 1	:B101.3
2	:Inspection, 12,1,2
3	:Overhaul, O.33, 1, 24
Box text	:Pump fails to start
Gate type	:P
Unavailability	:3.0E-03
Time indep unavail	:6.0E-04
Failure rate	.6.7E-06
Inspection interval	:30 days
Mean time to repair	:8 hours
Reference	:1.7
Comments	:
Component type	:Centr. pump/stand-by
Failure type	: 3
Data last changed by	:
Data last changed on	:

Handling Levels in SUPER-TREE   FIGURE 6

changing failure data for individual components, for groups of components, or classes of components, in order to perform a sensitivity analysis. Results are presented in bar chart and graphical form.

FRANTIC

The availability of stand-by systems varies with time since, in additional to the system components' failure probabilities, test intervals, and repairs of revealed failures, will contribute further to the picture. Thus FRANTIC creates an unavailability function for the system under analysis, based on the cut-set list generated by the CUTSET program. This function is compounded by individual component data from SUPER-TREE such as failure frequency, repair times and test intervals.

By providing lists and graphs of unavailability as a function of time, this program is useful for planning and evaluating the testing and maintenance of system components.

FRANTIC was originally developed by the US Nuclear Regulatory Commission.

SAMPLE

While the above programs all work from point values of failure probabilities, it is important to know the uncertainty distribution for key events. The unavailability function as generated from CUTSET by FRANTIC for the top event, is compounded from SUPER-TREE by details of distribution parameters for component failure rates and repair times, using the SAMPLE program. This program uses Monte-Carlo simulation to compute the uncertainty distribution for the top event.

SAMPLE was originally developed by the US Nuclear Regulatory Commission.

COST

The LCC of a plant comprises two basic components: the initial costs and the recurring costs. The COST program covers all aspects of these costs, some of which are specified by the user (interest rates, foreign exchange rates, etc.) some of which originate from SUPER-TREE (e.g. initial component costs, scheduled maintenance requirements) and some of which are generated by CUTSET (unavailability of the plant leading to production losses, corrective maintenance costs etc.).

A sensitivity analysis facility allows the LCC's dependence on key parameters to be assessed, in helping to keep costs at an optimum level with respect to safety and availability targets.

APPLICATIONS

The analysis and programs described in this paper are applicable to any plant or system, large or small. The advantages to be gained in analysing large systems include the handling of the following tasks:

- drawing and data-setting of the first complete set of fault trees

- updating of the fault trees as the analysis proceeds and changes are introduced.

- co-ordination of the work of several analysts contributing to the whole study.

FTA, as part of a safety assessment, on availability assessment or a LCC analysis finds applications in many branches of modern technology.

Some examples are:

Power generation
Power transmission and distribution
Telecommunications
Aerospace
Transport
Chemical process industries
Oil and gas production transmission and distribution
Marine systems

Retaining models of fault trees and LCC on file throughout the installation's operational life provides a powerful management tool in assisting with decisions involving design, equipment and organisational or monetary fluctuations.

REFERENCES

1. Watson I A, "Safety and Reliability Procedures in Various Industries". Safety and Reliability Directorate UKAEA, SRS/GR/76, January 1989.

2. Hirschberg S, and Knochenhauer M. "SUPER-NET, a Multi-purpose Tool for Reliability and Risk Assessment". International Post-SMIRT 10 Seminar. "The Role and Use of PCs in Probabilistic Safety assessment and Decision Making". Beverley Hills, California, August 21-22, 1989.

3. Björe S, Hirschberg S, and Knochenhauer M . "A Unified Approach to Reliability Analysis". Society of Reliability Engineers Symposium, Vasteras, Sweden, October 10-12, 1988.

4. Hirschberg S et al. "A Comparative Uncertainty and Sensitivity of an Accident Sequence" Ibid.

5. Knochenhauer M, Olsson L, and Alm S. "Verification of Availability Guarantees in HVDC Projects: Estimation and Optimisation of the Impact from Corrective and Preventive Maintenance". Reliability Achievement: The Commercial Incentive. SRE-Symposium, Stavanger, Norway, October 9-11, 1989.

ACKNOWLEDGEMENT

The author would like to thank ABB Atom, Västerås, Sweden, for permission to publish this paper.

INCORPORATING HUMAN FACTORS INTO FORMAL SAFETY ASSESSMENT: THE OFFSHORE SAFETY CASE

by

Linda J. Bellamy and Tim A.W. Geyer
Four Elements Limited

ABSTRACT

The undertaking of a Formal Safety Assessment or Safety Case provides an opportunity for the offshore industry to consider the human contribution to risk and to examine the Human Factors aspects of safety management. This paper summarises an approach for incorporating Human Factors into offshore QRA, including the evaluation of escape, evacuation and rescue. It then proceeds to describe a framework for evaluating the Safety Management System with respect to managing human error. The essential components of the SMS are identified as Demand Optimisation, Capacity Optimisation, Incentive Motivation, and Feedback Control.

INTRODUCTION

In 1979 the nuclear accident at Three Mile Island highlighted the human element as a major contributory cause (Kemeny, 1979). Control room design, communications, and management were central features in the failure of the system. The accident gave rise to a great flurry of activity on many aspects of human reliability and its assessment. Similarly, following the publication of the Piper Alpha Inquiry (Cullen, 1990), the need for a review of the potential applications of Human Factors (HF) to offshore systems is evident and has already resulted in an unprecedented interest in the subject.

This paper sets out an approach towards potential HF applications to Formal Safety Assessment. The purpose is to highlight those areas where a company or installation's control of human error could be demonstrated in the offshore Safety Case.

RATIONALE

Cassidy (1989) has indicated that one of the regular omissions in submitted Safety Cases is the implications of human reliability, and the inherent uncertainties. John Rimington (Director General, HSE) pointed out at the Piper Alpha Inquiry that recent major incidents had focused attention on the significance of the management of safety, "... the chain of command for safety, ... leadership from the top ... in-firm safety culture and particularly the influence of the human factor in accident causation ... The establishment of a safety culture included ... the systematic identification and assessment of hazards and the devising and exercise of preventive systems which are subject to audit and review. In such approaches particular attention is given to the investigation of error. The control of human error involves the assumption that people will make mistakes but that by thought, pre-design and proper motivation this can be made much more difficult and the consequences mitigated." (Cullen, 1990 p.357).

The nuclear and chemical industries, and their regulators, are increasingly concerned about the safety management factor which has now become an issue for Human Factors research in the UK. However, it is still recognised that many basic Human Factors principles, guidance and techniques are not easily incorporated into system design and management, although there are efforts to increase awareness and offer practical advice (Health and Safety Executive, 1989). Also, changes in technology bring about new problems. For example automation, which many regard as a means of eliminating the human problem, can have the effect of pushing the human error into other parts of the system such as design (Bellamy & Geyer, 1988). Under such conditions, the recipe for disaster is that the designer works on the basis that personnel are error free whilst personnel assume that the system will fail safe.

Human failures are exacerbated by inadequacies in design, information, procedures, inspection, supervision, and training. All too often, personnel are blamed for making errors where the potential for such errors could have been recovered by dealing with these inadequacies in the system. Wilful violations, such as taking short cuts, are more likely to arise where the "performance shaping factors" (PSFs) in the work context are poor, for example difficulties in carrying out the procedure, problems in access to equipment, inadequate tools, time or production pressures, or poor planning and preparation. Also, unanticipated operating or maintenance problems may give rise to the application of ad hoc procedures which have been inadequately reviewed for the human activities involved.

Poor performance shaping factors can be recovered before they have the effect of increasing the likelihood of error and the potential for disaster such as occurred on Piper Alpha. Failure to attend to such preventive or recovery mechanisms means that the potential for error will lay dormant within the system. Such 'latent failures' as they are called, are not easily visible. Yet they contribute to over 90% of accidents (e.g. Bellamy, Geyer and Astley, 1989).

It therefore makes sense to tackle not only the problem of identifying the human errors which may occur in a system, but also the underlying causes and the failure to prevent or recover them. This directs attention to Safety Management Systems. Cullen, in the recommendations arising from the Piper Alpha Inquiry (Cullen, 1990), puts emphasis on the assessment of the Safety Management System (SMS) as part of a Formal Safety Assessment or Safety Case. But how should such an assessment be carried out?

In this paper we develop a Human Factors approach to SMS assessment which is based on our understanding of the causes of accidents in hazardous industries. Firstly, we consider the Human Factors input to the other important areas emphasised by Cullen. These are Quantitative Risk Assessment (QRA) and the escape, evacuation and rescue of personnel in the event of an emergency.

HUMAN FACTORS IN QRA

One of the most successful penetrations of HF into risk assessment has been in the form of Human Reliability Assessment (HRA) where the technique for quantifying human error can be readily taken up by non-specialists. It is not our intention to discuss the details of the methods here (see Bellamy, Kirwan and Cox, 1986; SRD, 1988). There are two areas where a human reliability input is generally important in QRA:

- Where operator action in the man-machine system is critical to achieving system reliability.

- Where operator action contributes to preventing escalation of an incident.

HRA is used as a sub-model of the QRA. The overall logical framework for HRA is the same as for hardware orientated QRA, the key techniques being the utilisation of fault and event trees. The rationale should be that wherever fault trees are used for quantification, assessment of human error must also be considered. However, in FSA for offshore installations it is unlikely that fault trees will be required where historical data already exist. Historical failure data indicating failures of equipment components (e.g. frequencies of hydrocarbon leaks from valves) will include the human causal contribution and there would be little benefit in analysing this contribution in detail in the QRA.

In formulating event trees the mitigation of the consequences of an event may be dependent upon certain direct human actions such as manual contributions to ESD, blowdown, etc. Therefore the quantification of the human contribution to escalation is a necessary part of the assessment. Historical human error data is unlikely to be available, therefore the analyst may use judgement of probabilities based on generic data or an alternative HRA technique. Typical error probabilities are shown in Table 1. Error probabilities may need to be modified, perhaps by as much as a factor of 10, to take account of performance shaping factors in the design of support such as the man-machine interface. There are other interesting problems, such as the effect of the preceding sequence of events on error likelihood.

Table 1 Selected Generic Human Error Rates (after Hunns and Daniels, 1980)

ERROR TYPE	TYPE OF BEHAVIOUR	HUMAN ERROR PROBABILITY
1	Extraordinary errors of the type difficult to conceive how they could occur: stress free, powerful cues initiating for success.	10^{-5}
2	Error in regularly performed commonplace simple tasks with minimum stress.	10^{-4}
3	Errors of commission such as operating the wrong button or reading the wrong display. More complex task, less time available, some cues necessary.	10^{-3}
4	Errors of omission where dependence is placed on situation cues and memory. Complex, unfamiliar task with little feedback and some distractions.	10^{-2}
5	Highly complex task, considerable stress, little time to perform it.	10^{-1}
6	Process involving creative thinking, unfamiliar complex operation where time is short, stress is high.	10^{-1} to 1

When accident scenarios have been evaluated, the actions of personnel in escape and evacuation need to be quantified in order to estimate fatalities. Generally, the approach has been to consider typical personnel locations and the proportion of time spent in those locations in relation to tasks that they have to perform. Response times for mustering in a safe haven, decision times to evacuate and the likelihood of successful evacuation must all be calculated, perhaps with the assistance of models for generating the data. In general, the use of reasonable assumptions has predominated. However, there is potential here for more sophisticated analysis as the use of event trees is extremely limiting. Because of the complexity of human actions, detailed evaluation of escape, evacuation, rescue and associated emergency control activities has tended to be avoided in the past and is particularly difficult to model in terms of the interactions with an escalating event; the need for this has now been emphasised by the Piper Alpha disaster. Some of the necessary considerations for the future are dealt with in the next section.

Although we have only discussed the very limited incorporation of HRA into Formal Safety Assessment, this is not to say its use in QRA is not important for other applications. Engineering design studies (e.g. for lifeboats) or detailed reliability studies (e.g. ESD systems, simultaneous operations such as drilling and production, heavy lifting operations) have greatly benefited from an HRA contribution.

HUMAN FACTORS IN ESCAPE, EVACUATION & RESCUE

Both the Piper Alpha and the Alexander Kielland disasters highlight, to different degrees, problems of design, information, training, procedures, communication and decision making, in escape, evacuation and rescue in emergencies.

The evaluation of an installation's escape, evacuation and rescue (EE & R) system, and associated emergency control activities, should consider whether all the required human actions have been supported by the design and procedures. Will the capacities of people under stress will be able to meet the demands of an emergency? This evaluation has to be undertaken in both QRA and SMS assessment (the latter is discussed in the next section).

Quantification of EE & R should be based upon an evaluation of the success of possible sequences of human responses to a representative set of scenarios derived from QRA. There are three critical factors to be taken into account in the scenario assessments:

- Available time (e.g. 30 minutes to platform collapse)
- Available courses of action (e.g. escape alternatives)
- Response goals (e.g. get to TSR)

These factors will change throughout the development of a scenario. They will affect and be affected by human responses as well as the design context and scenario characteristics. For example, early recognition of an event will increase available escape and evacuation time.

For each scenario, fatalities would be dependent upon:

- Time available for EE & R
- Personnel locations/manning levels
- The likelihood of installation/emergency control/rescue personnel carrying out appropriate actions *in time*, including recognition, communication, and decision making
- Performance shaping factors (design characteristics, procedures, weather conditions, etc.)

The aim would be to model:

- Whether appropriate actions occur *in time* to avoid the threat (e.g. whether the platform is evacuated before structural collapse).
- Whether these actions are successful, given the inherently hazardous nature of certain of them (e.g. lifeboat evacuation, jumping into the sea). The hazards associated with certain actions will also vary, dependent upon factors such as weather and the capacities of personnel to meet action demands (e.g. problems in gripping ladders when wearing survival suits).

The quantification of actions occurring in time has been used successfully before, for example in the development of event trees for recovery from watchkeeping failure in ship-platform collision studies. However, ideally, to reduce the size of the analyst's task, and particularly to increase the potential of the analysis, modelling support is needed to generate and quantify the appropriate sequences without the constraints of a particular structure like an event tree.

The analysis should enable possible improvements in EE & R to be identified which could lead to reduction of fatalities. Such improvements are likely to relate to reduction in demands on personnel, such that these demands are well within the capacities of personnel to respond through better design support, and means of increasing available time such as through early recognition, and reduction in the total response time through well timed communications and decisions.

HUMAN FACTORS ASSESSMENT OF SAFETY MANAGEMENT SYSTEMS

Background
There are two ways of looking at the human factor in safety. One is what we will call *generic*, the other is company or *site specific*. The former focuses on activities or tasks associated with to certain operations or design, the latter focuses on safety management.

Consideration of the human element on a generic basis means determining how the "average" person or team would be expected to perform, given a particular procedure in the context of a particular design. Applications of human factors to the assessment of offshore installations has tended to be of this generic type (task analysis, HRA, HF reviews utilising design guidance).

The assessment of Safety Management Systems has rarely been addressed, not only in the offshore industry but also in other hazardous industries. Currently, there is a great deal of research interest in this area. For example, the Health & Safety Executive are sponsoring HF research into the development of modification of risk (MOR) techniques. Such techniques would utilise audits and models to evaluate and quantify the effects of management quality on the likelihood of failure and the mitigation of failure consequences and impact. This would be for interfacing with QRA, particularly for CIMAH sites (Hurst, Nussey and Pape 1989, Bellamy et al 1989). The rationale is that failure rates of equipment (pipework, vessels, etc.) and the effectiveness of accident mitigation measures (emergency control) are sensitive to safety management factors and that these factors can be quantified through an auditing process. This work is progressing through detailed accident analysis, audit method applications, and analysis of safety attitudes in industry.

However, such research programmes have long term goals. The need for offshore operators to present an assessment of their Safety Management Systems as a response to Cullen's recommendations is an urgent one. How can Human Factors help? From the perspective of the human factors specialist, the interest is directed towards any factors which may influence the performance of operations and maintenance personnel who interact directly with the system. Such influences range from "remote" causes such as economic/production pressures, to proximal ones like control room interfaces and escape route design. However, this has to be translated into a framework that enables companies to demonstrate, in a fairly flexible way, that the SMS is controlling human error.

This section discusses the assessment of the Safety Management System with respect to the management of human error. We present a simple framework for SMS assessment based on identification of the management controls which could have prevented the occurrence of accidents, coupled with the application of basic principles of human performance optimisation.

Objectives
The foremost Human Factors objectives of the safety management system (SMS) should be:

1.   To provide operating personnel with:

- a design that they do not have to fight;
- procedures which are not bureaucratically cumbersome, difficult to perform or hazardous;
- necessary and unambiguous information;
- an environment conducive to minimising stress and discomfort.

This is *Demand Optimisation*.

2.   To select and train personnel such that their knowledge and skills are appropriate to the tasks which they have to perform, and to design jobs so as to maximise personnel performance capacities, not reduce them. This is *Capacity Optimisation*.

3.   To motivate people to perform safely and to minimise pressure to do otherwise. This is *Incentive Motivation*.

4.   To monitor performance, identify deviations from safety standards, and to eliminate conditions conducive to error or procedure violation. This is *Feedback Control.*

Demand Optimisation

If a design makes excessive demands on personnel, the SMS will be considerably handicapped in achieving its objectives. Therefore, this part of the Human Factors assessment of the SMS should be to carry out a Human Factors review of the design demands.

For an operating installation, it would be a very sizeable task to review all the critical human activities by utilising 'generic' Human Factors techniques. HRA or task analysis, for example, are every time consuming. It is far better, therefore, to utilise 'generic' techniques, such as applying guidance, at appropriate stages in design and thereby minimise the occurrence of poor design features which could increase demands in a whole variety of activities. In this way, one reduces common mode human failures.

A basic question is, therefore, whether it can be demonstrated that such Human Factors considerations of demands were undertaken at the design stages of an installation and whether the management commitment to safety through demand optimisation is evident in the product. Human Factors reviews are therefore an important part of this demonstration, whether of documentation or in the form of a site audit of an operating installation, to collect objective 'evidence' that excessive demands on personnel have been minimised.

Capacity Optimisation

The subjective experience of demands is relative to the capacities of the personnel who have to respond to them. The balance of demands and capacities determines workload, both physical and mental, and if the balance is good then the likelihood of error and wilful violations will be reduced.

Appropriate selection and training for the tasks which personnel have to perform will maximise personnel performance capacities. Also, the organisation of different tasks into jobs will affect whether the best use of personnel capacity is achieved. This is evident in the fact that the possibilities for reduced manning have to be explored not only through reducing demands, but also in terms of training requirements for necessarily redesigned jobs.

The SMS should therefore be assessed in terms of the quality of its organisation and human resources i.e. the compatibility between the organisation of tasks into jobs, the required knowledge and skills, and the methods of selection and training.

Incentive Motivation

People have to be motivated to perform and to perform safely. The concept of motivation can be understood through consideration of external incentives or individual needs. If incentives are weak, or needs are fulfilled, motivation will be low. This would be the desirable state for unsafe acts. The importance of motivation should not be underestimated. Powerful needs or incentives can result in the denial of even the most convincing of information.

Assessment of the SMS in this respect should therefore concentrate on identifying what the incentives for safety are, and incentives for unsafe practices and whether these have been eliminated. Incentives may be positive (attractive) or negative (to be avoided). Typical incentives are those relating to physiological, psychological and social needs (eg. food, praise, money, friendship, pain, punishment, isolation, rejection).

Therefore, factors for investigation should include, for example, pay, team structures and personnel relationships, performance targets and associated rewards, personal development, peer group and other organisational pressures, disciplinary systems, accountability, job satisfaction, competing incentives (particularly production pressures).

Feedback Control
The effectiveness of the SMS can only be assessed in relation to the goals or standards which have been set on the basis of the safety policy of the organisation. These safety goals must be both achievable and laudable, and regularly reviewed in this respect.

Achievement of realistic safety goals must be, from the Human Factors perspective, through control of demands, capacities and incentives. These controls will require change if they are not working effectively. Continual feedback is therefore essential to making such adjustments. The way in which an organisation monitors the effectiveness of its controls, and uses that information to modify them to improve performance or reset its goals is therefore of vital importance to the SMS.

So, the existence of a safety policy and associated safety goals and standards must be established in the assessment of the SMS. It should be demonstrated that communication systems are in place for performance feedback and constructive comment from the lowest levels of the organisation up to the highest levels of management where the company's safety goals are set. It should be demonstrated that systems for regularly collecting feedback information are in place (eg. meetings, safety reviews, audits, incident and near miss reporting and investigation schemes). The quality of the information collected should be sufficient for identifying whether standards are being met, and whether there are any requirements for change.

The mechanism for feedback control must also be capable of reasonably rapid follow up where change is required. Limitations to follow-up action should be expressed in terms of what is reasonably practicable. Therefore, it is important to describe the climate in which the SMS operates, eg. economics, regulations, resource availability, industry norms and know-how.

SUMMARY

The potential for the utilisation of HF techniques is very wide, ranging from detailed task analysis to HF audits. For Human Factors to be effectively incorporated into FSA, the approach should be to demonstrate that human error is being controlled as far as possible. This means not only examining the tasks which people have to perform, particularly safety critical tasks, but also the underlying causes of error latent in the design and management of an installation.

REFERENCES

Bellamy, L.J. and Geyer, T.A.W. (1988) Addressing Human Factors Issues in the Safe Design and Operation of Computer Controlled Process Systems. pp. 189-202 in B.A. Sayers (Ed.) Human Factors and Decision Making, Their Influence on Safety and Reliability. London: Elsevier.

Bellamy, L.J., Geyer, T.A.W. and Astley, J.A. (1989). Evaluation of the Human contribution to Pipework and In-Line Equipment Failure Frequencies. HSE Contract Research Report No. 15/1989. UK Health and Safety Executive, Bootle, Merseyside.

Bellamy, L.J., Kirwan, B.I. and Cox, R.A. (1986) Incorporating Human Reliability into Probabilistic Risk Assessment, pp 6.1-6.20 in Proceedings of the 5th International Symposium, Loss Prevention and Safety Promotion in the Process Industries, Cannes, France, September 1986.

Cassidy, K. (1989) CIMAH Safety Cases: 1, Overview. pp. 220-232 in F.P. Lees and M.L. Ang (Eds.) Safety Cases, London: Butterworths.

Cullen, the Hon. Lord (1990) The Public Inquiry into the Piper Alpha Disaster. Department of Energy, 2 vols. London: HMSO.

Health and Safety Executive (1989) Human Factors in Industrial Safety. Health & Safety Series Booklet HS(G)48. London: HMSO.

Hunns, D. and Daniels, B.K. (1980) The Method of Paired Comparisons and the Results of the Paired Comparisons Consensus Exercise. Proceedings of the 6th Advances in Reliability Technology Symposium, vol. 1, pp. 31-71, NCSR R23, National Centre of Systems Reliability, Culcheth, Warrington.

Hurst, N.W., Nussey, C. and Pape, R.P. (1989) Development and Application of a risk Assessment Tool (RISKAT) in the Health and Safety Executive. Chemical Engineering Research and Design. 67(4), 362-372.

Kemeny, J. (1979) The Need for Change: The Legacy of TMI. Report of the President's Commission on the Accident at Three Mile Island. Washington, D.C.

Norwegian Public Reports (1981) The Alexander L. Kielland Accident. Report of the Commission appointed by Royal Decree of 28 March 1980 to Ministry of Justice and Police, March 1981, Norway.

SRD (1986) Human Reliability Assessors Guide. Edited by P. Humphreys. Publication RTS 88/954, Safety & Reliability Directorate, UK Atomic Energy Authority, Warrington.

FIRE RISK QUANTIFICATION USING A DISCRETE SCENARIO MODEL

Dr P M Thomas   (BNFL Engineering)
Dr J S Singh   (HEL Ltd)

ABSTRACT

A model has been developed to allow for quantification of fire risk in terms of heat and smoke exposure of individuals. Items such as fire and smoke spread, fire detection and alarm, and escape routes have been included. The model, known as SNARF (Systematic Numerical Assessment of the Risk of Fire) has been outlined and its use illustrated by application to a few simple hypothetical building layouts.

At present there is considerable research activity in the field of fire safety and numerous 'models' have been developed to quantify fire hazards. Some of the methods are very specific in scope, perhaps looking at a single variable during a fire, while others are extremely elaborate and need large mainframe computers. These techniques can all be useful in certain situations, but they do not address the overall fire risk problem.

There are, for example, methods for estimating the spread of smoke for a given fire condition. To make proper use of these methods it is important to look at the range of possible fires and their relative likelihood, and then combine these different aspects. To complete the fire study, the elements of fire behaviour have to be combined with the response of people in terms of their influence on the fire and their ability to escape.

Current building design practice relies heavily on safety standards and codes of practice. However, these codes cannot always be applied - for example, in cases of buildings with very low fire loads or where fire compartments are unusually large. In these circumstances an alternative approach to fire safety design needs to be considered. Such cases may include laboratories or other industrial premises.

The technique described here, known as SNARF (Systematic Numerical Analysis of the Risk of Fire), addresses these issues and provides a means for quantifying the overall fire risk, taking account of combustion, structural design, human behaviour and probabilistic elements. It is original in concept and scope.

It is being developed for BNFL Engineering by HEL Limited, as part of a programme to establish improved methods for fire safety analysis of new building projects.

The Discrete Scenario Concept

There are a large number of ways in which a fire can develop after initiation, depending on the physical circumstances at the time of the fire. In theory, every fire is capable of growing large enough to engulf an entire building or stopping immediately after initiation. In fact, most fires will be of some intermediate size.

A review of the statistical data on fire losses shows that one of the main features determining the losses from a fire is the time at which the fire is first detected. Early detection of fires frequently leads to low losses. Another important feature in fire protection is the presence of fire barriers. These can limit the extent of spread and thereby reduce the level of damage.

Taking a pragmatic approach, it is not necessary to consider every possible type of fire in every situation. The total number of possible incidents can be reduced to a manageable level by averaging appropriate variables. The loss of sensitivity and accuracy due to such simplification can still be kept within the bounds of the input statistical data (for example fire frequency data).

In the present model, two possibilities for detection time are considered. For each time, three possible conditions of fire barriers are considered:

- barrier open;

- barrier closed but fails;

- barrier closed and remains intact.

These possibilities define the scenarios which are quantified. The methodology traces each discrete scenario and evaluates the consequences in terms of the manner of spread and the consequences of death/injury. Based on the building design and the fire characteristics, the frequency of each scenario is also calculated. These two items of information - frequency and consequence - are then combined to give the risk from each scenario. Summation of the risk from each scenario over an entire building gives the risk associated with the building.

The output from the calculation is the number of fatalities per year which may be expected. This result can be directly compared with historical fire statistics. The number is not just an indicator of hazard based on judgement or experience (which is the outcome from 'points' methods for example) but is an absolute measure of the hazard.

The model is constructed in a manner that allows detailed analysis of the components making up the final risk. One feature is that it is possible to identify the risk contribution from different compartments and then consider how that outcome has occurred. The availability of a breakdown provides, among other things, guidance on the most efficient means for fire risk reduction. Since the model is based on an evaluation of the combustion (fire) and building design parameters, it is possible to evaluate the influence of important variables in detail.

Structure of the Model

Risk calculations using the model follow a precisely defined procedure (illustrated in Figure 1) using sub-models and other data. The structure allows easy extension of the model if necessary. Three sets of documentation are provided:

-    a step-by-step structured calculation procedure;

-    data-sheets for tables of data and analytical equations;

-    worksheets for data entry and record of calculation results.

Calculation Procedure

The calculation procedure has been formalised into a series of 8 worksheets (WS) shown in Figure 2. these worksheets call upon a standard database and library of fire safety models which have been tailored to a common format and simplified where necessary. A brief description of the worksheets follows to illustrate the scope of the model:

Worksheets 1 and 2 All the data relating to the building and its design is entered on these sheets prior to the analysis.

Worksheet 3 Fire development parameters are determined for each source compartment and include fire duration, time to flashover, barrier failure and smoke filling time.

Worksheet 4 Considers fire (flame) and smoke spread from each source compartment to all neighbouring (target) compartments.

Worksheet 5 Uses smoke spread information to estimate the time at which fire from each source compartment may be detected.

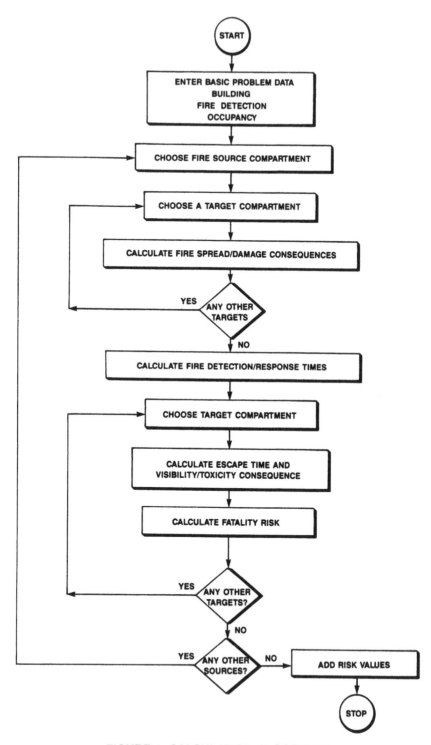

FIGURE 1: CALCULATION ALGORITHM

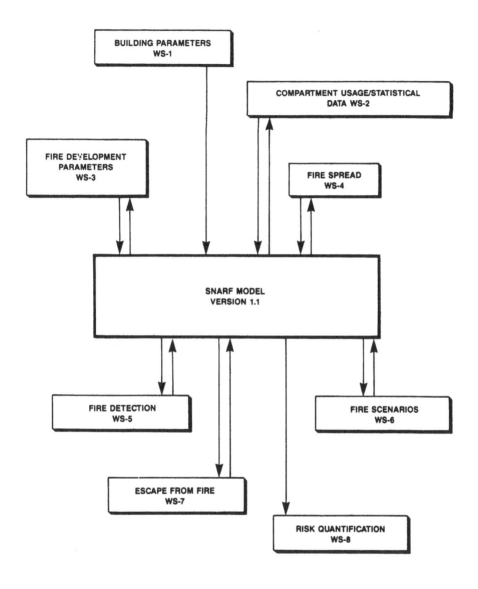

FIGURE 2: INFORMATION FLOW BETWEEN WORKSHEETS

Worksheet 6 Collates the information on fire and smoke spread and determines the toxicity, visibility and health aspects.

Worksheet 7 Evaluates the evacuation process for all scenarios, examining all escape routes and determining the impact of the fire as people attempt to escape.

Worksheet 8 Combines the various frequency and consequence parameters to evaluate risk for each compartment in turn which is then summed.

Defining the Problem

The value of the risk model is best illustrated by means of a practical example. The plan of the building used for the illustration is shown in Figure 3.

There are five compartments, numbered 1 to 5, and fire doorways, A to E. The only exit to the outside is through A and the hallway, compartment 3. The barriers C, D and E are of 60 minute fire resistance, B of 30 and A is an ordinary door (15 minute resistance assumed). All barriers are left open for between 5 and 20 per cent of the time. There are no automatic fire detectors anywhere in the building. The compartment use and occupancy is shown in Table 1.

To appreciate the value of the results, the above building should be reviewed and the following general questions considered:

-    What is the likely overall level of fatality risk as compared with the average for the UK?

-    Which compartment or design feature is likely to dominate the risk (and why)?

-    What single protection feature is likely to have a major impact on reducing the risk?

Base Case Fire Fatality Risk

Application of the calculation procedure to the above design shows that the total risk due to fire is 9.2×10^{-4} y^{-1} per person at risk. Compared with the fire general risk for published statistics typically of the order 10^{-6} y^{-1} per person this is very high.

In terms of the people at risk, the relative contributions to fatality are shown in Figure 4. This shows compartment 2 is the largest contribution to risk, closely followed by compartments 4 and 5 - ie. the people in these compartments are more at risk. The contribution from compartment 3 is negligible.

Another way to compare hazards is by source of risk, that is in terms of the compartments where fire is initiated. This is shown in Figure 5.

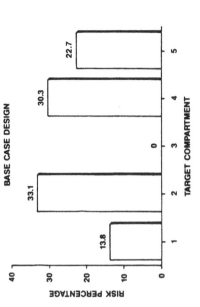

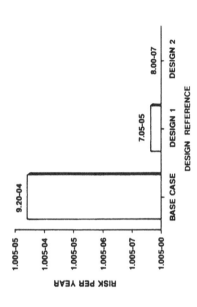

FIGURE 4: CONTRIBUTION TO RISK AS TARGET

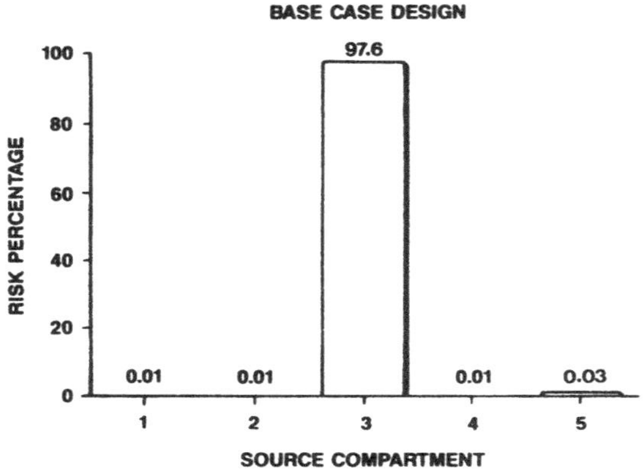

FIGURE 6: RISK COMPARISON FOR 3 DESIGNS (PER PERSON)

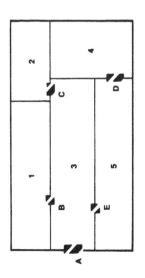

FIGURE 3: PLAN OF THE BUILDING

FIGURE 5: CONTRIBUTION TO RISK AT SOURCE

Compartment	Average Occupancy (Number)	Fire frequency $(y^{-1} \times 10^4)$	Fire load MJ/m^2
1 Store	1.5	2.18	600
2 Office	3.75	1.69	100
3 Hallway	1.8	4.58	500
4 Consultation	1.6	6.06	1800
5 Meeting room	5.0	6.20	-

Table 1 Compartment Use and Occupancy

The result in this case is unexpected at first the risk is dominated by fire starting in compartment 3. There are two overriding conclusions that emerge from the base-case building design:

- The fatality risk is more than two orders of magnitude higher than the historical average and therefore the design should be improved.

- Fires originating in compartment 3 are the major cause of fatality.

Discussion of the Base Case Risk Results

The brief analysis of the results clearly shows that to reduce the risk it is necessary to understand the reason for dominance by incidents originating in compartment 3. A look at the building layout shows the primary cause: fires in this compartment trap occupants in the rest of the building. This is due to the location of the fire exit - it can only be reached through compartment 3. Therefore people in the compartments furthest from the exit are most at risk. An analysis of the risk breakdown by target reveals the risk per person is highest in compartments 2 and 4. These are furthest from the fire exit.

There are two generic ways to tackle this problem: either provide alternative exits for the remote compartments or reduce the risk at source. One reason for people being trapped is that if a fire starts in compartment 3 when there is no one present, the smoke will build up to high levels until it reaches other areas and is detected. Therefore, it would be interesting to look at the effect of more rapid detection based on the same building design.

Design Alternative 1: Automatic Detection

It is now possible to use the models to evaluate the influence of the design change on the risk. Thus, the second example considers the early detection of fires in compartment 3 by installation of an automatic smoke detector.

Repeating the risk calculations, the model shows that the individual risk for the new design is reduced by over 90 per cent to 7.05×10^{-5} per year.

The risk breakdown in terms of the source of the fire and the targets is given in Table 2:

Compartment	Contribution to risk (%)	
	Source	Target
1	2.7	<0.01
2	1.2	3.8
3	10.0	3.0
4	<0.01	92.7
5	86.0	0.5

Table 2. Contribution to risk for design alternative 1

Clearly, in the new safer design, the risk is dominated by fires originating in compartment 5. the previous dominant source, compartment 3, is still significant since fire detectors are not foolproof. In terms of contribution from targets (ie. the people actually at risk), compartment 4 is the most vulnerable, even though as a source of fires it was quite minor.

The fire risk for the new design incorporating the smoke detector leads to the following conclusions:

-    The risk is dominated by fires that begin in compartment 5.

-    The occupants in compartment 4 are most threatened.

The reason for this rather surprising result again becomes obvious from consideration of the building layout. Occupants in compartment 4 must go through 5 in order to reach a safe exit. Therefore, whenever fire detection in compartment 5 is delayed, it presents escape problems for occupants in compartment 4.

Notice that in some instances, conclusions about fire safety can be reached simply from a review of the designs based on experience. However, experience will not give an absolute measure of hazard, nor will it indicate the extent of improvement following a design change. Similarly, the relative benefits of different options cannot easily be assessed by judgement alone.

Design alternative 2: Different Escape Route

Since the problem has been isolated to escape from compartment 4, it is appropriate to look at this aspect. The most direct alternative is to change the position of exit D which presently connects compartment 4 to 5 and provides the means of escape. If this is moved so it connects compartment 4 to 3 directly, then fires in compartment 5 should have little influence on occupants of compartment 4. This design change has been evaluated quantitatively using the model.

Compartment	Contribution to risk (%)	
	Source	Target
1	42.6	28.6
2	<0.01	29.6
3	57.4	13.3
4	<0.01	18.7
5	<0.01	9.8

Table 3. Distribution of fire risk for design alternative 2

The total risk for the design with a new position for the exit (and the fire detector in place as before) on an individual basis in equal to 8×10^{-7} y^{-1}. This is now in line with the average annual fatality rate from fires and represents an acceptable design. A comparison of the total risk values (per person) for the three designs is shown in Figure 6 (on a logarithmic scale).

The risk breakdown for the present design alternative in terms of fire sources and targets is given in Table 3.

There is now a much more event distribution of risk in terms of targets, although as sources, compartments 3 and 1 are dominant.

Detailed Analysis of Risk Figures

At this point it is useful to look in more detail at the way in which the risk comes about for the final design. The methodology allows this in a very simple manner.

Table 4 shows the breakdown of risk according to the time to fire detection: primary (early) or secondary (late) detection, for the two dominant sources. Clearly late detection dominates the fatality risk. This is in line with national fire statistics.

Fire Source Compartment	Contribution to total risk (%)	
	Early Detection	Late Detection
1	9.5	33.1
3	<0.01	57.4
TOTAL	9.5	90.5

Table 4. Effect of fire detection time on risk (design alternative 2)

It is now possible to look in more detail at the late detection incidents - say from compartment 3 which is the major source. The 57.4 per cent of the total risk - due entirely from this source - can be broken down in terms of the manner in which the fire spreads. In this case, it is possible to consider the contributions in terms of the condition of the barrier (fire door) from compartment 3: left open deliberately, left closed but fails due to fire heat or remains intact throughout the fire. The contributions from these three possible situations are shown in Table 5.

State of Fire Door	Contribution to risk (%)
Deliberately open	2.6
Closed but fails	15.9
Closed and survives	81.5

Table 5. Effect of fire barrier on risk (design alternative 2)

This shows that contrary to popular opinion, the presence of the fire door in a closed state leads to an increased risk of fire. This is true for this particular building layout and should not be regarded as a general rule. The presence of the door delays detection and allows the build-up of smoke. Then, when the fire is finally detected, building occupants are forced to travel through the smoke-filled source compartment as this is the only escape route.

It should be noted that the presence of fire barriers would have a positive influence in terms of restricting the growth of the fire. Structural safety would be improved.

Conclusions

A systematic fire risk model has been established which enables the hazard to be calculated directly for various building configurations. It contains all the parameters considered relevant to the safety of personnel in buildings. Using a simple example it has been demonstrated that the model can be used to determine the most effective fire safety design. Moreover, the structured nature of the method allows the contribution of each compartment of fire risk to be determined.

Its application could ultimately be widespread, for example it may be used to compare the fire risk associated with alternative building layouts, or examine the relative benefits of automatic detection and fixed fire fighting. The ability to meaningfully compare alternatives - passive and active protection for example - should be particularly valuable.

SESSION E:OPERATIONS AND OPERATIONAL SAFETY

CONTINGENCY   PLANNING   TECHNIQUES

TO   REDUCE   RESPONSE   TIME

FOR   TACKLING   OFFSHORE   WELL   BLOWOUTS

by

KEN   FRASER

NORTH   SEA   WELL   CONTROL   ENGINEERING   LTD

Summary

This paper discusses Contingency Planning Methods which can be applied to predict the nature and correct response for potential Offshore Well Blowouts. The paper lists the procedures to be followed to collate relevant data and how they can be applied to reduce response times.

Index

1.   Causes of Blowouts on Offshore Rigs.

2.   Effective response prediction and management.

3.   Application of collected data to reduce emergency response time.

1.1   Causes of Blowouts on Offshore Rigs

1.0   Introduction

Formation pressures are controlled during normal drilling and workover operations by means of the hydrostatic head exerted by the column of fluid in the well. This hydrostatic head is manipulated by adjusting the drilling or workover fluid density.

Such pressure control is termed Primary Control.

If Primary Control is lost in the well then Blowout Preventors, Surface Valves and /or Tubing Valves are used to provide a means of closing in the well to trap pressures and prevent well flow whilst a denser fluid is circulated into the well bore to regain Primary Control.

The use of this equipment is termed Secondary control..

During Production Operations, Primary Control is removed to allow the well to flow. Consequently, only Secondary Control can be used during Production. This is usually in the form of Surface and Downhole Safety Valves and a Xmas Tree.

Only if both Primary and Secondary well control is lost can a Blowout occur. The fact that they are lost does not mean that a Blowout will occur - only that the conditions then exist which would allow Blowouts to happen. Whether or not there will be a Blowout will depend on the exposed formations, their fluids and potential productivity.

For the purposes of this paper the rig operations during which a Blowout could occur are treated in 3 ways:

    1.   During drilling
    2.   During workover
    3.   During production operations.

1.1   Blowouts during Drilling Operations

1.1.0   Introduction

Any porous and permeable formation will try to produce its contained fluid into the Wellbore once Primary Control has been lost. Providing Primary Control is maintained

ie P Hydrostatic Head of Mud > P Formation

the well will not flow. When the well formations begin to flow, the well is termed to be kicking or 'taking a kick'

All kicks occur therefore when Primary Control is lost.

The 'kick' is the precursor to the Blowout during drilling operations.

<u>Causes of Kicks during Drilling</u>

Since drilling is a dynamic process, the potential causes of a kick at any time are directly associated with the activity in progress at that time.

The Drilling Process can be generally considered in 5 modes.

1.   Tripping in the hole
2.   Drilling ahead
3.   Making connections
4 .   Circulating
5.   Tripping out of the hole

1.1.1   <u>Causes of Kicks while Tripping in the Hole</u>

If the trip out of the hole has been completed without incident and the well is static during BHA handling then it is likely that Primary Control is effective. When we run in the hole we can affect this Primary Control in two ways:

a)   Pressure surging creating losses
b)   Stringing out of swabbed in gas

1.1.1a)   Pressure Surging

That is by running in the hole the surge pressure caused by lowering the string is sufficient to break a weak formation causing a loss of hydrostatic head in the annulus. This loss of hydrostatic head can allow previously controlled formations to flow into the wellbore.

NB. It is loss in hydrostatic head which causes the kick and not the pressure surging itself. Providing that the hole is kept full then Primary Control will be maintained even with induced losses down hole.

1.1b)   Stringing Out of Swabbed in Gas

When tripping out of the hole it is possible to swab in small amounts of gas as the BHA passes reservoir rocks without a clear indication at surface. Normally this phenomena exhibits itself in the form of TRIP GAS.

Trip Gas is a common occurrence on wells and indicates how easy it is to swab wells in. What is worth commenting on is how trip gas is usually recorded at 'bottoms up' circulation. This indicates that the swabbed in gas has stayed in the same spot in the wellbore during the Round Trip or if it has migrated, it has not migrated very far.

The migration speed of gas in mud is often quoted as being 1000'/hour however in the Author's view this figure is meaningless since the actual migration speed depends on influx make up, solubility, formation porosity, mud type, lithology and bore hole geometry.

All that can be said with any certainly is that the influx will not sink, it will either stay put or ascend.

If the influx ascends then it expands as the hydrostatic head exerted on it reduces. As it expands it should become detectable, however this depends on the sensitivity of the Rig Instrumentation and the vigilance of the crew.

Since most muds are engineered with a TRIP MARGIN built into them a small amount of swabbed in trip gas expansion will not immediately allow the well to kick, however it will obviously reduce the overbalance.

If there is a gas influx sitting in the Open Hole occupying a height of say y ft in 12 1/4 Open Hole and then a 8" BHA is run into it the height of the influx will become $\pm$ 2y ft.

This obviously reduces the hydrostatic head on the bottom of the hole and could be enough to allow a new influx to enter the well bore.

1.1.2 Causes of Kicks Whilst Drilling Ahead

Kicks occur whilst drilling ahead for three reasons:

a)   Penetrating a higher pressured zone.
b)   Penetrating a loss zone and losing Primary Control.
c)   Losing Primary Control due to gas cutting.

1.1.2a)   Penetrating a Higher Pressured Zone

If a reservoir type rock is penetrated whilst drilling ahead which has a higher pressure than can be primarily controlled by the drilling mud then the well can kick.

1.1.2b)   Penetrating a Loss Zone

If a new formation is drilled which has insufficient strength to support the hydrostatic head of a full column of mud to surface then the fluid level in the well will fall until it reaches an equilibrium point. At this point the hydrostatic head exerted by the mud will balance the formation strength of the newly drilled weaker formation.

If there are any reservoir type rocks higher up the hole which have a formation pressure higher than the new hydrostatic head exerted by the shorter mud column then the well can kick.

1.1.2c)   Losing Primary Control due to Gas Cutting

As rock which contains gas in its pore spaces is drilled, this gas is circulated towards the surface together with the drilled cuttings. As it rises the hydrostatic head which is compressing it is reducing and

As the gas expands it occupies a greater percentage of the annular volume hence reducing the hydrostatic head of mud below it.

Clearly any amount of released gas from cuttings will to some degree reduce the hydrostatic head of the annular mud column, however small amounts will probably have such a minimal effect that the TRIP MARGIN of overbalance in the mud will be sufficient to ensure that the potentially productive formations downhole are still subject to full Primary Control.

If however a large amount of cuttings are released (usually by high penetrating rates) and/or   the gas content of the rock is high then Primary Control can be lost and the well allowed to kick.

1.1.3 Causes of Kicks Whilst Making Connections

Whilst pulling the bit off bottom to make a connection there is a swabbing tendency, however since in most cases the pumps are running as the bit is pulled back thus swabbing tendency is negated.

The principle cause of kicks whilst making a connection is the loss of the effect of the ECD on the bottom of the hole when the pumps are turned off to add a single.

If a higher pressured formation has been encountered whilst drilling the single down it is possible that the additional hydraulic head exerted on the bottom of the hole due to circulating pressure losses in the annulus might have been sufficient to maintain Primary Control.
When the pumps are turned off the hydrostatic head at the bottom of the well is reduced and in some cases can prove to be insufficient to provide Primary Control over the new formation.

In such cases the well can kick.

1.1.4 Causes of Kicks Whilst Circulating

A well can not be brought into production by the circulating process alone so any kicks which are taken during circulation were probably created earlier by another mechanism. The circulation itself would merely be bringing the kick to surface where it becomes detectable.

1.1.5 Causes of Kicks Whilst Tripping out of the Hole

There are two mechanisms whereby the well might kick during a trip out of the hole.

These are:

a)   due to swabbing
b)   if the hole was not circulated clear prior to the trip then a higher pressure zone which had just been drilled into prior to the trip could make its presence felt.

1.1.5a)   Swabbing

This is the most common cause of Blow Outs. If the hole is dirty, the BHA packed off with cuttings, the Trip Margin in the mud too small or the tripping speed is too fast for the conditions then the well can be swabbed in and a kick taken. The reason why so many wells are lost due to Blow Outs created by swabbing is that the influx that can be swabbed in can be very large. Furthermore, since the kick detection does not take place until later on in the trip this implies that the bit is some distance off bottom. This situation restricts our well killing options.

1.1.5 b)   Undetected higher pressured zones

If a formation which has high pressures but low permeability is drilled into just prior to a round trip, it is possible that Primary Control can be lost. This situation, however, might not be immediately obvious due to the low flow rates from the producing formation. It is partly for this reason that normal safe drilling practice is to circulate bottoms up prior to round tripping so that any influx can be detected. In this way, the kick is detected with the drill string at depth allowing recovery of Primary Control in the quickest and most effective manner.

1.1.6   Summary

Statistically, in Drilling Operations, the most likely cause of kicks which lead to Blowouts is swabbing. Comparatively few rigs or platforms are lost by kicks caused by other mechanisms since the kicks thus taken tend to be smaller and with the drill string on bottom also tend to be easier to control.

When wells Blowout the wells will either produce through a failed or inoperative BOP system or alternatively will produce up the outside of the casing string.

1.2   Blowouts during Workover Operations

The primary well control considerations which apply to drilling operations also apply during Workovers.

Workover fluids of known density are used to maintain Primary Control on exposed productive formations during workovers.

Since production casing has been set and completion tubing design always includes the facility to set valves to hold pressure within it, Kicks and Blowouts should not occur during Workovers. In practice, however, this is not the case and as many Blowouts occur during Workovers as during Drilling Operations.

The causes of such Blowouts are in most cases poor programming, poor Workover practices and generally a lack of appreciation of what could go wrong.

When wells Blowout during Workover Operations they tend to produce through a failed BOP or Xmas tree at surface.

1.3   Blowouts during Production Operations

To gain production from a well, Primary Control must first be lost. The hydrostatic head of fluid in the tubing must be less than the producing formation pressure otherwise the well will not flow.

Due to this situation, all controls on wells in production are 'secondary' controls:

> Xmas trees
> Surface safety valves
> Downhole safety valves

Blowouts during production operations are therefore due to failures in the secondary controls and imply that the well will produce through a failed Xmas Tree or flow line.

1.4   Summary of how wells blow out - narrative

When wells blow out in most cases, they will produce through damaged, missing or inoperative surface equipment. Only in the case of drilling operations can we expect the well to produce from the seabed through the sea either:

a)   around the casing, or
b)   when a subsea BOP system fails and allows the well to flow into the sea below the rig.

Conclusions to be drawn

By defining the operations to be carried out, the manner in which the well might blow out can also be defined.

Having defined this, we must ensure that any Response Plan must meet the needs of the particular operations.

1.5 <u>SUMMARY OF HOW WELLS WILL BLOW OUT – PICTORIAL</u>

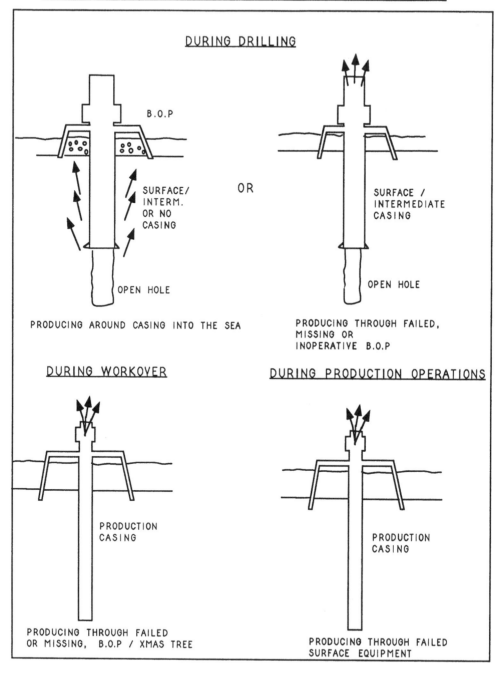

2.   Effective Response Prediction and Management.

The initial questions which have to be faced in the event of a Blowout are:-

Q1   Can the well be capped?
Q2   Will the well flow quit as the hole bridges due to Formation Collapse?
Q3   Will a relief well be necessary and if so where should it be drilled from?
Q4   What are the implications of the Blowout?

Q1   Can the well be capped?

In all cases, except when the well is producing from around the casing, it is feasible that the well could be capped. Just how feasible, will depend on the particular well and its surface environment. Suitable criteria to be considered when assessing this are:

      a.   Production Potential.
      b.   Knock on effects on installation and other wells.
      c.   Effectiveness of deluge systems.

Q2   Will the well flow quit as the hole bridges?

To predict this we need to know something about the producing formation. In the North Sea most formations characteristics are well documented and to a great degree - predictable. Most surface formations tend to be unconsolidated and will ultimately collapse and bridge whilst deeper lying formations will be able to withstand open hole production rates more readily without bridging for longer periods. To resolve this question, we must determine:

      d.   How stable are the producing formations.

Q3   Will a relief well be necessary and if so where should it be drilled from?

In practice, most Blowouts are stopped and repaired by relief wells. A relief well is drilled into the original well bore and by pumping fluids at high rates down the relief well, Primary Control is restored.

The type of Rig to be used for relief wells depends on the following criteria:

      e.   Water depth.
      f.   Sea bed type and obstructions.
      g.   Pressure rating requirements for relief well rig.

Positioning the rig is dependent upon.

      h.   Depth of Blowout.
      i.   Prevailing wind and waves direction and speeds.
    (f)   Sea bed type and obstructions.

Q4   What are the implications of the Blowout?

To assess this, the following criteria must be considered:

        (a)   Production potential
        j.   Potential of toxic gas production
        k.   Potential of oil spill
        l.   Potential new formation pressure regimes
        m.   Proximity to shipping lanes.

To summarise, in order to efficiently respond to a Blowout, we need to know the following.

        a.   Production potential
        b.   Knock on effects on installation and other wells
        c.   Effectiveness of deluge systems
        d.   How stable are the producing formations.
        e.   Water depth
        f.   Sea bed type and obstructions
        g.   Pressure rating requirements for relief well rig.
        h.   Depth of Blowout
        i.   Prevailing wind and waves direction and speeds
        j.   Potential of toxic gas production
        k.   Potential of oil spill
        l.   Potential new formation pressure regimes
        m.   Proximity to shipping lanes

Some of this information will be available whilst the rest can be predicted with some accuracy. By considering all of the above criteria prior to drilling a well the likely control scenario that would be required in the event of a Blowout.

The mechanism whereby this can be achieved in practice is listed as below

a.   <u>Production potential.</u>

By talking to the Petroleum Engineers who worked on raising the Drilling prospect, an estimate of the exposed formations production potential can be made. This estimate should include details of the anticipated formation fluids and the rate at which they could be expected to flow in an uncontrolled manner.

b.   <u>Knock on effects on installation and other wells.</u>

This can be estimated by carrying out a HAZAN survey on the wellhead area and its surroundings.

c.   <u>Effectiveness of deluge systems.</u>

This too is estimated using HAZAN techniques.

d. How stable are the producing formations.

This can be predicted by the Petroleum Engineers who worked on the drilling prospect.

e. Water depth.

These are known in advance of any drilling operation commencing.

f. Sea bed type and obstructions

This information is also known prior to siting the first rig at the location.

g. Pressure rating requirements for relief well rig.

This can be predicted from the estimate of formation pressures to be encountered in the initial well.

h. Depth of Blowout

The formations capable of producing in a manner capable of causing a Blowout can be predicted with reasonable accuracy by the Petroleum Engineers.

i. Prevailing wind and waves direction and speeds

This is known for each area.

j. Potential of toxic gas production

This can be predicted by the Reservoir Engineers and Production Geologists.

k. Potential of oil spill

This can be predicted for each potential producing horizon by the Petroleum Engineers.

l. Potential new formation pressure regimes

This can be interpolated from Petroleum Engineering data and lithology.

m. Proximity to shipping lanes

This is known for each area.

3. **Application of collected data to reduce emergency response time for tackling an offshore Blowout.**

Having collected the above date, we are now in a position to make the following preparations.

1. Relief Well Siting

A provisional relief well site can be selected conforming to the following criteria:

Ensuring that no geological faulting might allow a Blowout to communicate with the chosen relief well site.

Relatively straightforward access by conventional directional drilling into original well bore.

Sufficient clear sea bed to allow anchoring of the drilling vessel in an acceptable manner.

2. Relief Well Rig Selection

The type of rig required for the relief well drilling can be specified as follows:

Type	Semi submersible or jack-up (usually delineated by water depth criteria).
Pressure Rating	Determined by anticipated downhole pressure.
Rig Heading	By studying wind and weather data the desired rig heading can be selected.

By considering the above, a detailed rig specification can be drawn up well ahead of any Blowout so that when a Blowout does occur the choice of rigs which could be used for drilling relief wells is apparent. The location that the rig will use to drill the relief well from has been specified and the rigs heading during relief well drilling is also known.

3. Relief well casing design requirements

By studying the original wells lithological column and anticipating Blowout surcharging of formations a provisional relief well casing design can be made. This invariably involves more casing strings than the original Well Design. By considering this requirement potential logistic problems in sourcing unusual casing sizes or specifications can be highlighted or overcome.

Summary

By using straightforward procedures, even before an Exploration, Appraisal or Development well is drilled, Contingency Plans can be drawn up to respond to potential Blowouts.

The application of the methods discussed in this paper allow the Relief Well Rig type and rating to be specified in advance.

A Relief Well location can be selected which satisfies Ergonomic, Environmental and Technical requirements for the potential Blowout.

By using this information, response time in a real emergency can be reduced dramatically.

By knowing the required Relief Well Rig specification an immediate assessment of availability of suitable Rigs can be made. The chosen Rig can be quickly mobilised to the chosen location and moored up on the current heading.

Furthermore, by knowing the potential casing requirements, immediate sourcing of non standard equipment can be implemented.

In practice, by using the above method, the Blowout Response time will be reduced by hours and in some cases days.

The impact of this Application of Technology includes cost savings Environmental Impact limitations and as such is an exercise worth carrying out on every well or platform.

UPSTREAM SAFETY PROGRAMME IN PETRONAS

IR. HAJI AHMAD NORDEEN DATO' SALLEH*

INTRODUCTION

PETRONAS is the National Oil Company of Malaysia established in 1974 to serve as the Government instrument in charge of petroleum matters and to exercise, on behalf of the country, its sovereign right over the country's petroleum resources. Upstream safety in the context of this paper refers to the safety of offshore petroleum activities in our exploration and production operations both for oil and gas resources.

We have just completed 16 years of successful management in the petroleum business. In order to continue with our success, we need to maintain the highest level of safety consciousness in the upstream sector both in the project and operational phases. The sector should be in a state of readiness and have the capability to handle any emergency situation relating to offshore petroleum activities.

---

* Group Chief Engineer, Safety and Technical Audit
  Marine and Safety Department
  Exploration and Production Division
  PETRONAS
  MALAYSIA

BACKGROUND

Crude oil and petroleum products, inclusive of LNG, are potentially hazardous by nature. In this light, safety in our petroleum industry is extremely important for our business survival, especially in upstream offshore operations where the environment is hostile, physical movement is rather restricted and the weather conditions are occasionally unfavorable and problematic. In addition, offshore installations are at a considerable distance from shore-based support.

Historically, the E & P sector of the oil industry had employed relatively low technology in its drilling and production of oil and gas. However, from the mid 1970s, we have seen considerable changes in the degree of complexity of the designs employed in the industry as well as in its approach to safety. In view of this, we need to continually keep abreast of these changes especially as a result of the Piper Alpha Disaster, by improving our safety programmes and the way we manage our upstream operations.

The upstream safety programmes can be categorized into three major areas :

(a) Technical Safety Audit and Risk Assessment

(b) Operational Safety and Health

(c) Regulatory and Emergency Response Preparedness

Please refer to Attachment I which provides a complete picture of how we monitor and regulate offshore operations by our Production Sharing Contractors (P.S Contractors). For your information, the PS Contractors are major oil companies

ATTACHMENT 1

FACILITY / DESIGN SAFETY

- STANDARDISATION OF PLATFORM SAFETY SYSTEM
- PRE-RELINQUISHMENT AUDIT
- FIRE PREVENTION TASK FORCE
- PARTICIPATION IN TECHNICAL SAFETY AUDIT
- SAFETY AND ALARM SYSTEM UPGRADE
- SPECIAL STUDY / SAFETY TASK FORCE
- PROJECT SAFETY REVIEWS

REGULATORY AND EMERGENCY PREPAREDNESS

- UPGRADING PETRONAS PROCEDURES
- CHECKING ADEQUACY AND CONSISTENCY OF PSC'S SAFETY PRACTICES
- ESTABLISHING EMERGENCY RESPONSE GUIDELINES
- GAUGING LEVELS OF EMERGENCY RESPONSE PREPAREDNESS

UPSTREAM SAFETY ACTIVITIES

PERSONAL SAFETY

- QUARTERLY SAFETY MEETING
- UPDATING SAFETY DATABASE
- SAFETY TRAINING STANDARDISATION
- ACCIDENT INVESTIGATION AND ANALYSIS
- SAFETY ALERT BULLETIN

- SAFETY CAMPAIGN AND AWARD PROMOTION
- CONTRACTORS SAFETY PROGRAMS ACCEPTABILITY
- HAZARD HUNTS
- SAFETY REPORTING
- OCCUPATIONAL HEALTH CONTROL

contracted by us to manage offshore operations on our behalf for a possibly 15-year period. There are four (4) of them currently operating and we have signed twenty eight (28) new Production Sharing Contracts since 1987.

MALAYSIAN REGULATORY REGIME

Currently Malaysia does not have complete statutory requirements of its own governing upstream petroleum activities. PETRONAS, on behalf of the Malaysian Government, has developed and established a number of procedures, which are considered minimum requirements for Production Sharing Contractors to adhere to and abide by. They constitute the following:

(1) General Procedures for Safety and Welfare (GPSW)

(2) PETRONAS Guidelines for Inspection and Maintenance (PGIM)

(3) Procedures for Drilling Operations (PDO)

(4) Procedures for Production Operation (PPO)

(5) PETRONAS Upstream Emergency Procedure Manual

(6) PETRONAS Risk Assessment Manual

The GPSW for example provides guidelines for PS Contractors' preparation of their own Safety and Occupational Health Procedures/Manual; while the PDO and PPO provide guidelines outlining PETRONAS' requirements on oil and gas drillings and operations respectively.

The above procedures are being reviewed and revised as and when necessary. The overall picture of our upstream regulatory requirements is as per Attachments II & III.

UPSTREAM REGULATORY REQUIREMENTS

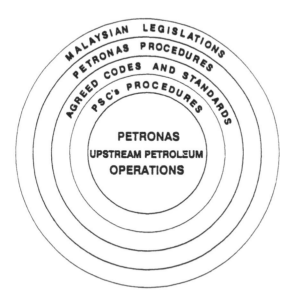

MALAYSIAN LEGISLATIONS

- Exclusive Economic Zone Act 1984
- Factories & Machinery Act 1967
- Petroleum (Safety Measures) Act 1984
- Environmental Quality Act 1974
- Continental Shelf Act 1966 (revised 1972)
- Petroleum (Income Tax) Act 1967
- Atomic Energy Licencing Act 1984
- Protected Area and Protected Place Act 1959
- Malaysian Merchant Shipping Ordinance

AGREED CODES AND STANDARDS

- API - American Petroleum INstitute
- ANSI - American National Standards Institute
- ASME - American Society of Mechanical Engineer
- ASTM - American Society of Testing & Materials
- AWS - American Welding Society
- NFPA - National Fire Protection Association
- IEEE - Institute of Electrical & Electronics Engineer
- BS - British Standards
- IMO - International Maritime Organisation
- ICS - International Chamber of Shipping
- OCIMF - Oil Companies International Marine Forum
- Classification Societies Rules

PETRONAS PROCEDURES

- GPSW - General Procedures for
    Safety and Welfare
- PPO - Procedures for Production Operations
- PDO - Procedures for Drilling Operations

PRODUCTION SHARING CONTRACTORS (PSC's) PROCEDURES

- Design Instruction Manuals
- General Fabrication Specifications
- Standard Engineering Specifications
- Safety Manuals
- Operating Manuals
- Maintenance Manuals
- Emergency Response Plan Manuals

PETRONAS PROCEDURES for EXPLORATION & PRODUCTION ACTIVITIES

NO.	MAIN PROCEDURES	SUPPLEMENTARY PROCEDURES
1.	General Procedures for Safety and Welfare (GPSW)	a) PETRONAS Upstream Emergency Communications/ Notification Procedure Manual b) Procedures for Diving Operations c) PETRONAS Guideline for Barges Operating Offshore Malaysia d) PETRONAS Emergency Response Guidelines for Exploration
2.	Procedures for Production Operations (PPO)	a) PETRONAS Guidelines for Inspection and Maintenance of Production Facilities b) Guidelines for Gas Measurement c) Guidelines for Dynamic Liquid Hydrocarbon Measurement
3.	Procedures for Drilling Operations (PDO)	a) Guide to Reporting and Sampling Requirements for Exploration Drilling

Regulations are another vital safety area of concern in our upstream operations. We need to regulate and control PS Contractors operations with regard to safety and occupational health by referring to all relevant codes, standards and procedures enforced (be it Malaysian, PETRONAS or internationally accepted).

TECHNICAL SAFETY AUDIT AND RISK ASSESSMENT

We make it a point in general to ensure all inherent safety issues related to upstream platform design are considered thoroughly. This takes into account current safety standards pertaining to offshore platform design, together with our Piper Alpha - safety task force findings and recommendations. We would insist that for a new platform design, HAZOPs should be conducted taking into account all possible operating deviations, preferably to be done by a third party. During the course of the projects, a series of safety reviews are carried out and followed thoroughly at various stages of the project development, to ensure the new platform built is as safe as is reasonably possible. In addition, we should conduct regular technical audit reviews to identify any deficiency in existing platform facilities, especially the ones to be relinquished. For instance, we are currently reviewing seven(7) offshore projects up to 1993, particularly in the following specialised areas :

(i)   Plant Layout and Area Classification

(ii)   Fire Water Requirement

(iii) Fire and Gas Detection System

(vi)   Process Isolations and Shutdown

(v)   Depressurisation, Blowdown and Venting

(vi)   Alarm System

(vii) Drainage and Waste Disposal System

(viii) Emergency Shutdown Valve (ESD) System.

In addition we have identified six(6) major fields whose platforms will be relinquished to PETRONAS in 1993 and 1995. This requires us to conduct pre-relinquishment technical safety audit - detailed enough to ensure these facilities are kept in safe and efficient production condition when we take over.

Attachment IV illustrates the overall scope of the Loss Prevention we hope to accomplish at various stages of our offshore activities. We continue to give top priority to Technical Safety Audit since there had been a number of major accidents in the last decade.

IMPACT OF PIPER DISASTER ON OUR INDUSTRY

The disaster has made us review and upgrade our loss prevention standard for our upstream operations - utilising the Department of Energy (DOE)'s Interim Report - Sept. 1988 and lately Public Inquiry Report by Lord Cullen - Nov. 1990.

LOSS PREVENTION ACTIVITIES

CONCEPTUAL — DETAIL DESIGN — MATERIALS & PROCUREMENTS — FABRICATION & CONSTRUCTION — INST & COMM — PSC PRODUCING OPERATIONS — POST PSC PRODUCING — ABANDON

SAFETY PHILOSOPHY REVIEW
RISK ASSESSMENT
HAZOPS

HAZOPS
RISK ASSESSMENT

SAFETY EQUIPMENT SPECIFICATIONS
SAFETY EQUIPMENT F.A.T

QA/QC PROGRAMME AUDIT
INSPECTION WORKSITES

COMMISSIONING AUDIT
PRE START-UP AUDIT

TECHNICAL SAFETY AUDIT — 4 yrs cycle
PROCEDURAL AUDIT — Once every 4 yrs (All PSC)
FACILITY CHECK(INSPECTION) — For identical facility/audit follow-up
DAMAGE EVALUATION — Major property damage
ACCIDENT/INCIDENT INVESTIGATION
LESSONS LEARNED — Worldwide

PRE RENUNISHMENT AUDIT — 3 yrs prior to relinquishment

TECHNICAL SAFETY AUDIT
PROCEDURAL AUDIT
FACILITY CHECK(INSPECTION)
DAMAGE EVALUATION
ACCIDENT/INCIDENT INVESTIGATION
LESSONS LEARNED

PRE ABANDONMENT REVIEW — 1 yr prior to abandonment

PERIODIC CHECK — Undismantled platform

We immediately reviewed all our manned offshore facilities during the period December 1988 to January 1989 in the light of the lessons learnt. Scope of Work covered (see Attachment V) are as follows:-

*   Safe Evacuation and Emergency Response
    - to look at our emergency response plan, living quarters, assembly areas, evacuation routes, procedures, training and drills.
*   Hydrocarbon Inventories and Exposure
    - to study platform and pipeline inventories, ESD system isolation, blowdown and exposure to safe areas.
*   Fire Protection and Safety
    - to look at ESD System design and reliability, fire protection system and equipment, alarm and communications systems and emergency lighting.
*   Operations and Procedures
    - to evaluate work permit procedures, removal of equipment from service, non-standard operations, shift and view change over and familiarization of new personnel.

Task force's findings and recommendations are evaluated. Some of the recommendations have been implemented immediately, whilst many of the more complex and high cost items have to be evaluated further prior to approval for implementation. Todate we have made a firm commitment to improve our installed facilities in the tune of

SCOPE OF WORK

o **SAFE EVACUATION AND EMERGENCY RESPONSE**

  - EMERGENCY RESPONSE PLAN
  - QUARTERS AND ASSEMBLY AREAS
  - EVACUATION ROUTES, PROCEDURES, TRAINING AND DRILLS

o **HYDROCARBON INVENTORIES AND EXPOSURE**

  - PLATFORM AND PIPELINE INVENTORIES
  - ESD SYSTEM ISOLATION, BLOWDOWN
  - EXPOSURE TO SAFE AREAS

o **FIRE PROTECTION AND SAFETY**

  - ESD SYSTEM DESIGN AND RELIABILITY
  - FIRE PROTECTION SYSTEMS AND EQUIPMENT
  - ALARM AND COMMUNICATIONS SYSTEMS
  - EMERGENCY LIGHTING

o **OPERATIONS AND PROCEDURES**

  - WORK PERMIT PROCEDURES
  - REMOVAL OF EQUIPMENT FROM SERVICE
  - NON-STANDARD OPERATIONS
  - SHIFT AND CREW CHANGE-OVER
  - FAMILIARIZATION OF NEW PERSONNEL

M$200 millions, that is about £35 millions

Quantitative Risk Assessment (QRA) has now played a major role in our safety assessment of new projects and modification projects. We believe it would help us to provide a useful tool in our decisions making. For example, we have currently been using QRA in establishing firewater requirements, platform layout improvements, subsea isolation requirements and passive fire protection.

PETRONAS is currently reviewing those recommendations made in the Cullen's report. Roughly we have estimated that:

(a) At least 50% are either in existence in our safety systems or have been implemented soon after the disaster.

(b) 25% are principally in-line with our present practices, and

(c) 25% are under considerations for further evaluation.

In summary, the Piper disaster do have tremendous impact to our Malaysian Offshore Industry. We no longer look at safety as the way we used to. We believe we now need to emphasise that :

   (i) Our PS Contractors are to set more realistic safety objectives and goals, and their plans to accomplish them.

(ii)   All projects, modifications and operations must demonstrate that they are safe through formal safety assessment, with all risks taken into considerations.

(iii)   The stage at which an assessment is conducted should be such that any resultant findings/recommendations can be incorporated either into the design or prior to a particular operation.

(iv)   Emergency response capability must cater for all risks and the full spectrum of their consequences.

(v)   More specific safety assessment period to be identified (either 3 or 5 year cycle) and conducted by the PS Contractors in their respective areas of responsibility.

All these activities would be monitored and controlled by PETRONAS.

SAFETY IN OFFSHORE OPERATIONS

In our Production Sharing Contracts with oil companies operating in our territorial waters, we make it clear that they are to conduct their petroleum operations in a safe and efficient manner. Safety depends on not only good engineering practices of constructing and operating the installations from which the drilling and production processes are conducted and on which the workforce is accommodated, but also good safety management and maintenance of safe systems of work. These operations involve personnel

with widely different skills and their safety and health
offshore are of paramount importance.

Towards achieving zero lost time accident (LTA) and to
minimize all types of incidents, we monitor every accident/
incident investigation report. This is to ensure basic
causes are fully identified so that correct preventive
measures can be taken to prevent a recurrence. If the
accident is due to the PS Contractor's negligence, we can
withhold approval of damage claims.

Please see Attachment VI,VII & VIII showing the 1990
Contractors Safety Statistics monitored. We measure their
safety performances by considering LTA Frequency Rate and
Severity Rate.

We meet quarterly with every operating PS Contractor, for
example to discuss various existing safety problems during
the previous quarter, and to consider what has been or could
be done to resolve them. We also review every quarterly with
the PS Contractors the status of their safety performance and
programmes. Attachment IX illustrates PS Contractors 1990
Programmes. We insist that their programmes should be as
realistic as far as possible.

To ensure safety is given priority in all phases of

1990 SAFETY PERFORMANCE IN PETRONAS OFFSHORE OPERATIONS

COMBINED – PERSONNEL AND CONTRACTORS

	Manhours Worked			LTA			Days Lost			Frequency Rate			Severity Rate		
	A	B	C	A	B	C	A	B	C	A	B	C	A	B	C
Quarter 1	2,077,533	3,583,309	959,269	2	2	1	29	74	4	1.0	0.6	1.0	14.0	20.7	4.2
Quarter 2	2,485,355	3,895,983	777,283	1	4	0	5	131	0	0.4	1.0	0.0	2.0	33.5	0.0
Quarter 3	2,752,396	4,288,353	1,125,377	1	4	4	33	103	116	0.4	0.9	3.5	12.0	24.0	103.1
Quarter 4	2,185,392	4,395,085	1,417,484	0	5	0	0	147	0	0.0	1.3	0.0	0.0	36.9	0.0
Year to Date	9,500,676	16,162,730	4,279,413	4	15	5	67	455	120	0.4	0.9	1.2	7.1	28.2	28.0
TOTAL	29,942,819			24			642			0.8			21.4		
				(52)			(923)			(2.0)			(35.4)		

YEAR TO DATE FIGURES

	Manhours Worked			LTA			Days Lost			Frequency Rate			Severity Rate		
	A	B	C	A	B	C	A	B	C	A	B	C	A	B	C
Personnel	3,147,514	5,759,880	2,624,805	1	2	1	33	69	4	0.3	0.3	0.4	10.5	12.0	1.5
Contractor	5,644,187	11,325,154	1,654,608	3	13	4	34	399	116	0.5	1.1	2.4	6.0	35.2	70.1

NOTE : ( ) are 1989 figures
LTA – Lost Time Accident

DISTRIBUTION OF LOST TIME ACCIDENTS
BY AREA OF ACTIVITIES

YEAR : 1989

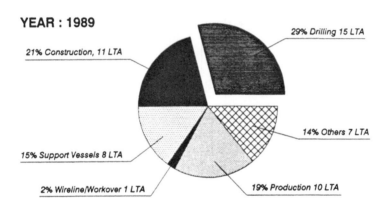

29% Drilling 15 LTA

21% Construction, 11 LTA

14% Others 7 LTA

15% Support Vessels 8 LTA

2% Wireline/Workover 1 LTA

19% Production 10 LTA

YEAR : 1990

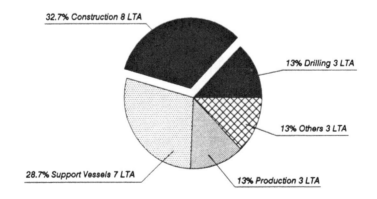

32.7% Construction 8 LTA

13% Drilling 3 LTA

13% Others 3 LTA

28.7% Support Vessels 7 LTA

13% Production 3 LTA

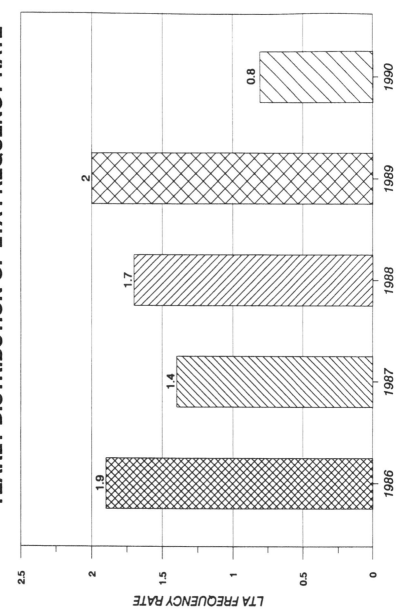

YEARLY DISTRIBUTION OF LTA FREQUENCY RATE

petroleum operations, we perform safety inspections at PS
Contractors' facilities. Any deficiencies are recorded in
accordance with their priorities for implementation. Our
findings are then discussed thoroughly with the relevant
operating staff before a follow-up action report is issued to
the particular PS Contractor.

Furthermore, we gather and analyse all accident/incident
reports in upstream operations, since they provide a
useful basis for evaluating the safety record and performance
of each operating PS Contractor. This includes property
damage or non-LTA incident. All incident/accident cases,
together with inspection findings, are regularly updated in
our computerized safety information system database.
Attachment VII shows distribution of LTAs by area of
activities. This analysis provides us with a useful
indication of the area we need to emphasise in our safety
management.

To enhance safety awareness among operators offshore, we also
issue a 'Safety Alert Bulletin' to our PS Contractors by
directly relaying relevant safety messages and
incident/accident cases which have a bearing on their safe
operations. Every offshore platform (currently we have 22
manned platforms) will be informed bimonthly on our safety
news.

Two significant safety activities for which we can claim success in our effort to upgrade safety operations are as follows:

   (i)   Centralizing safety training schools for offshore operating staff and subcontractors. We are currently standardizing all safety training courses both in East and Peninsular Malaysia. We emphasise effective and standardized safety training courses as a way to upgrade relevant safety skills of both the contractor and subcontractor workers since such training is vital for accident prevention.

(ii)   Standardizing safety systems and equipment installed by PS Contractors so as to achieve effective emergency response. Identified areas are :

   (a)   Manned platform alarm system

   (b)   Terminal alarm system

   (c)   Fire and gas detection system

   (d)   Piping colour coding system

What have we achieved in implementing all these safety programmes in our upstream operations? Attachment X illustrates our achievements of LTA Frequency Rate : Malaysian PS Contractors Vs Worldwide Oil/Gas Industry

Even if our performance is better than the global rate, we cannot afford to be complacent. We will continue to

PS CONTRACTORS 1990 SAFETY PROGRAM

	PS CONTRACTOR A	PS CONTRACTOR B
1990 SAFETY TARGET	• Limit LTA to:16 Contractors and 1 Personnel. • No injury goal.	• Zero fatality. • 30% reduction in LTA Freq. Rate for combined Contractors and employees.
SAFETY AWARENESS PROGRAMS	• Off the job safety - Conduct defensive driving seminar. • Enhanced Contractors Safety Awareness. • Implement "One Day At Time" program. • Implement Injury Prevention Programs.	• Encourage Contractor's Management on Important of Safety Awareness. • Issue of new Safety & Environmental Policy leaflets. • Safety video, publication and posters. • Unsafe Act Auditing. • Campaigns competion & quiz
SAFETY TRAINING	• Emphasis on Contractors safety training program. • Establish Contractor run Safety & Survival Training similar to SRIBIMA Training School.	• Safety passport program. • Conduct enhanced defensive driving course. Complete fire fighting/prevention course to Auxiliary Firefighting Team and extend to Firemen. • Refresher course on Unsafe Act Auditing. • Emergency Response Training • Continue COMPANY/SRIBIMA Training.
EMERGENCY RESPONSE PREPAREDNESS	• Enhanced Emergency Response Capability. • Conduct major drills.	• Continue Emergency Response • Team Training with more exercise and live drills. • Centralised Emergency Response Procedures on site specific basis.

113

MALAYSIAN vs WORLDWIDE OIL/GAS INDUSTRY

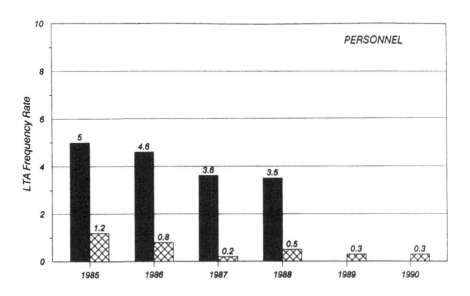

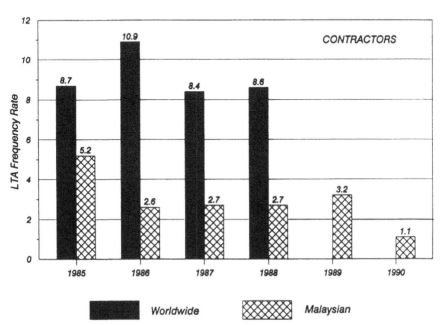

maintain or further improve our record on a year-to-year basis.

EMERGENCY PREPAREDNESS

In the area of Emergency Response (ER) Exercise/Drill preparedness, we develop PETRONAS ER's guidelines including checklists, to (a) review new PS Contractors' plan/procedure manual, (b) plan and implement 'ER' exercises/drill programmes and (c) gauge the level of 'ER' preparedness of each PS Contractor's facilities. Attachment XI illustrates our Emergency Response Preparedness:

While the main emphasis is on the prevention of accidents, we place equal importance on the PS Contractors' readiness and capability to handle any emergency situation associated with our upstream operations. They are to ensure their respective ER procedures are in order and drill exercises on major and minor accident scenarios conducted regularly, aimed at minimum losses. We normally participate in the exercises and provide our recommendation(s) for improvement.

As for oil spill contingency and emergency response plan is concerned, PETRONAS and its PS Contractors have focussed a lot of efforts in preventing oil spill in line with PETRONAS target of zero oil spill occurance. We have not experienced any spill of the same magnitude as Valdez as shown by the oil

EMERGENCY RESPONSE PREPAREDNESS

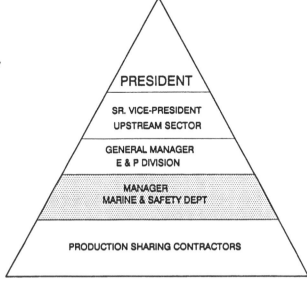

- **Accountability**

  PRESIDENT

  SR. VICE-PRESIDENT
  UPSTREAM SECTOR

  GENERAL MANAGER
  E & P DIVISION

  MANAGER
  MARINE & SAFETY DEPT

  PRODUCTION SHARING CONTRACTORS

- **Goals**

  EMERGENCY RESPONSE PREPAREDNESS: ALL ARE IN ORDER
  - MANUALS
  - EXERCISES

  ACTUAL EMERGENCY : ALL ARE UNDER CONTROL
  (If it does occur)
  AIM FOR - MINIMAL LOSS OF LIFE
  - MINIMAL PROPERTY LOSS

- **PETRONAS Roles**

  - UPGRADE AND REVIEW EMERGENCY RESPONSE PLAN MANUALS
  - GAUGE THE LEVEL OF EMERGENCY RESPONSE EXERCISES
  - ADVISE TOP MANAGEMENT OF EMERGENCY SITUATION

- **Production Sharing Contractors Roles**

  - COMMITTED TO ACHIEVE THE ABOVE GOALS

spill incidence records from 1987-1990 (Attachment XII).
Despite the absence of a catastrophic spill, Production
Sharing Contractors are required to submit individual oil
spill contigency and response plan, before conducting
drilling and production operations as per both PETRONAS
Procedures for Drilling and Production Operations. All of
them comply with this requirement. In terms of oil spill
capability, most of the PS Contractors which are currently in
the exploration stage, comply with PETRONAS minimum
requirement for chemical dispersion capability during
offshore exploration drilling. This minimum requirement does
not stop exploration contractors, in having additional
containment and recovery capability as well as being an
associate member of the Tiered Area Response Capability
(TARC) stockpile in Singapore.

Among the producing Contractors, Sarawak Shell Berhad is
currently in the process of upgrading its current oil spill
capability of 10,000 barrels. While Esso Production Malaysia
Inc. through its tiered response capability concept
(Attachment XIII) is confident of their capability after
conducting 2 major exercises in 1990 in combating a 70,000
barrels spill.

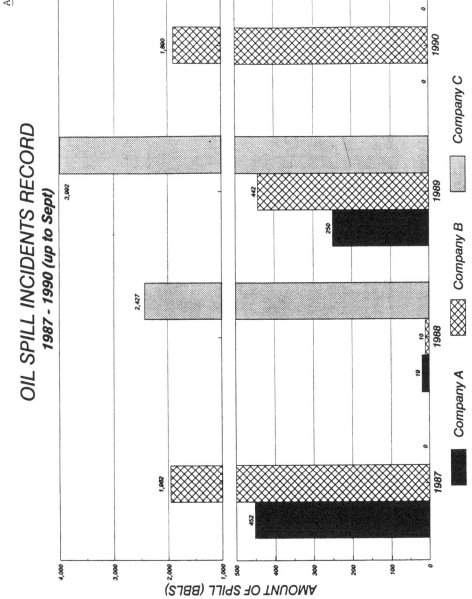

OIL SPILL INCIDENTS RECORD
1987 - 1990 (up to Sept)

TIERED RESPONSE CAPABILITY CONCEPT

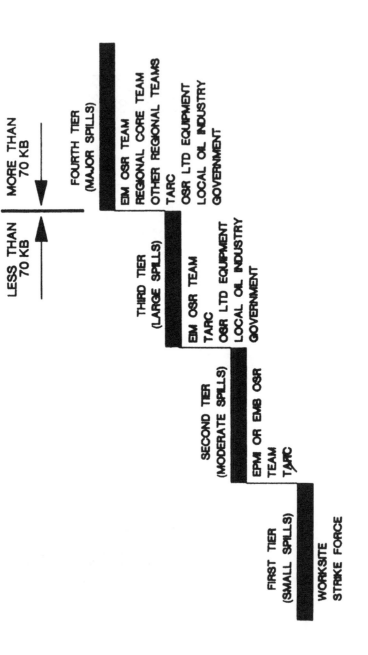

LESS THAN 70 KB

MORE THAN 70 KB

FIRST TIER
(SMALL SPILLS)

WORKSITE
STRIKE FORCE

SECOND TIER
(MODERATE SPILLS)

EPMI OR EMB OSR
TEAM
TARC

THIRD TIER
(LARGE SPILLS)

EM OSR TEAM
TARC
OSR LTD EQUIPMENT
LOCAL OIL INDUSTRY
GOVERNMENT

FOURTH TIER
(MAJOR SPILLS)

EM OSR TEAM
REGIONAL CORE TEAM
OTHER REGIONAL TEAMS
TARC
OSR LTD EQUIPMENT
LOCAL OIL INDUSTRY
GOVERNMENT

SAFETY AND TECHNICAL AUDIT GROUP

To carry out the above three major programmes effectively and
efficiently, the PETRONAS upstream sector has now revamped
its organisation structure whereby inhouse control of PS
Contractors within our Exploration and Production Division
is now being placed under Marine & Safety Department - Safety
& Technical Audit Group. The Group is responsible and
accountable for interfacing with all the PS Contractors.
It is manned by engineers with 10 - 20 years of working
experience in the petroleum industry.

CONCLUSION

We believe safety improvement is a never-ending process. Our
motto is to tackle all identified safety problems in
upstream operations as quickly as possible before any major
accident happens. Secondly, whenever there is a disaster
involving upstream operations, we would react immediately
to get to the underlying causes and undertake the
necessary remedial actions.

Since the objectives of productivity, quality and cost
control are intrinsically related to safety at work, we are
constantly monitor and react immediately to any safety issues
associated with our upstream operations.

To conclude, we would need to (a) ensure high quality in

safety design and engineering (b) enforce practical safety techniques and procedures (c) motivate constantly all operating staff and contract/subcontract workers to work safely. This way, we would be contributing towards the success of PETRONAS in managing national oil resources, in line with modern safety practices.

Thank you.

THE MARINE CHEMIST PROGRAM:
NFPA AND INDUSTRY AT WORK TO PROMOTE MARITIME SAFETY

G. R. COLONNA, NATIONAL FIRE PROTECTION ASSOCIATION, USA

ABSTRACT

The Marine Chemist program represents a unique
partnership between industry and government that originated
nearly seventy years ago. The Marine Chemist performs a
service essential to the overall safety of the entire maritime
industry. He anchors a program that recognizes, evaluates
and controls health, fire and explosion hazards associated
with the repair and construction of marine vessels. His
primary objective is to preserve and promote safety to life, limb
and property within the marine industry. He is now
recognized as a specialist in confined space entry safety and
the related atmospheric testing or sampling. Today's program
serves as a model for cooperation between government and
industry and showcases the Marine Chemist as a multi-
talented safety and health professional.

INTRODUCTION

The National Fire Protection Association (NFPA) marks its involvement
with the maritime industry almost as far back as its own development. The
NFPA is a private, non-profit, consensus standards-making membership
organization. The NFPA mission dedicates the Association to making
people safe from fire and related hazards through education and
development of technically based fire protection codes and standards. Not
only does NFPA develop marine fire protection standards, it also
administers the training and certification program for the professional
charged with the program implementation. The Marine Chemist program
is unique in its origin as well as its implementation.

MARINE FIELD SERVICE PROGRAM

The Marine Chemist program traces its origins to shortly after the conclusion of World War I. It was at that time that representatives of the marine insurance industry and vessel owners approached NFPA to request assistance with the fire and explosion problems which were experienced in US shipyards during and immediately after the war. NFPA responded by developing Appendix A of the NFPA Marine Regulations. This document contained recommendations of safe practices to be followed when repairing vessels. It became the forerunner of NFPA 306, the Standard for the Control of Gas Hazards on Vessels, which today must be followed by all Marine Chemists. The unique aspect of this early standard was that it not only specified requirements to be followed in the repair and construction of marine vessels, but it also required a specially trained individual to implement those requirements. This individual, in 1922, became the first Marine Chemist. At that time, the American Bureau of Shipping (ABS) directed the program for the certification of Marine Chemists; they retained that responsibility until 1963.

Just as industry had created the program and the standard safety requirements, it again took the lead in expanding the program to include a more formalized procedure for certificating Marine Chemists and also for providing technical field service support to the individual Marine Chemists. In 1963 the National Fire Protection Association agreed to take over the program for certification of Marine Chemists from the ABS and to work with the industry to expand the program. Industry made this request to NFPA and agreed to fund the development of a broader program that would be capable of providing Marine Chemists with both the administrative support, as well as the technical support. This request was made by the American Bureau of Shipping, American Hull Insurance Syndicate, American Institute of Merchant Shipping, American Petroleum Institute, American Waterways Operators, Hull and Cargo Surveyors, Marine Chemist Association and Shipbuilders Council of America. Today, these same organizations comprise the membership of the NFPA Marine Field Service Advisory Committee.

Since 1963, the NFPA Marine Field Service Program's most visible activity has been the training and certification of Marine Chemists. The Marine Chemist is the specially qualified professional identified in the original Appendix A of the NFPA Marine Regulations. The industry helped create the safety requirements, created the Marine Chemist and funds the varied aspects of the program through collection of a surcharge attached to the price of every Marine Chemist inspection certificate issued. Today, the program returns some of the industry contribution back to the industry in the form of training projects. More will be discussed on this topic later in the paper. First, it is important for you to have a more complete understanding about the Marine Chemist; who he is, why he exists, and how he operates.

NFPA CERTIFICATED MARINE CHEMIST

What is a Marine Chemist?

Marine Chemists are trained professionals whose responsibility is to ensure that repair and construction of marine vessels can be made in safety whenever a potential exists that those repairs can result in a fire or explosion. By virtue of their day-to-day activities, Marine Chemists are uniquely qualified and trained as specialists in confined space safety and atmospheric sampling or monitoring.

The Marine Chemist was created in 1922 and today is certificated by the National Fire Protection Association. There are about one hundred certificated Marine Chemists actively practicing within the industry today. The program that the industry created nearly seventy years ago remains essentially intact and the same as when it began. One significant change is that the government now recognizes the Marine Chemist in the various safety regulations of the Occupational Safety and Health Administration (OSHA) and U.S. Coast Guard (USCG).

The qualifications and performance of Marine Chemists are attended to by the Marine Chemist Qualification Board which consists of representatives from the marine insurance industry, shipyards, tankship operators, Marine Chemists and government (OSHA, USCG, Navy). These members and those of the Marine Field Service Advisory Committee are appointed by the NFPA Board of Directors. Both activities advise the President of NFPA and its Board of Directors on the program budget and new project ideas.

What are the qualifications of a Marine Chemist?

The Rules for Certification (approved by the Board of Directors) require that all Marine Chemists:

    o Have at least a 2 year technical degree

    o Complete 6 college level chemistry courses

    o Have work experience in both the marine industry and an analytical or petrochemical laboratory

    o Complete a twenty-one module training curriculum that includes basic fire and explosion chemistry, shipyard safety, confined space safety, tank cleaning, hazardous chemicals properties, instrumentation, and industrial hygiene

    o Complete at least 300 hours of on-the-job inspection training

    o Complete a written exam on the NFPA 306 standard, government regulations and industrial hygiene

    o Complete an interview with the Marine Chemist Qualification Board

Once certificated, a Marine Chemist is required to apply for recertification every five years. Conditions for recertification include a minimum level of activity, written examination, review by the Marine Chemist Qualification Board and attendance at professional training seminars.

What are the duties and responsibilities of a Marine Chemist?

The Marine Chemist provides key safety and health guidance and services to the marine industry. His duties largely involve the recognition, evaluation and control of the hazards that are primarily associated with entry and work in confined spaces on marine vessels that are undergoing contruction or repair. The Marine Chemist and the maritime regulations and standards comprise a very practical and effective systems approach devoted to preserving and promoting safety to life, limb and property within the maritime construction and repair industry. Though the program originated as a precaution against fire and explosion hazards within the repair activity, today the increased awareness of health <u>and</u> safety hazards has necessitated action by the Marine Chemist to protect workers from toxic hazards as well as fire and explosion.

As part of the maritime confined space safety program the Marine Chemist provides the following:

Recognition of the spaces which pose a hazard to workers during entry or work as a result of difficulty with entry and egress, unnatural ventilation, and/or the presence of or potential for introduction of any atmospheric contaminants.

> The spaces are typically cargo and fuel tanks, void spaces, ballast tanks, pump rooms, and in some cases, engine rooms or machinery spaces. These spaces may be deficient in oxygen, contain flammable or toxic contaminants or residual liquid or scale that is capable of regenerating hazardous conditions. These spaces must be opened and entered for inspection prior to a change of cargo and for repairs. The Marine Chemist in accordance with his training and experience often provides guidance to shipyards concerning the preparation of the space for worker entry on such aspects of the work as cleaning and ventilation.

The Marine Chemist identifies the spaces and the potential hazards that may affect worker safety. To determine the actual presence of atmospheric contaminants and their severity, it is necessary for the Marine Chemist to Evaluate the hazards.

Evaluation involves real-time monitoring of the confined spaces to determine the atmospheric levels of oxygen, flammables, and toxics within the spaces. The Marine Chemist compares the results of his real-time monitoring or measurements with guidelines provided in the standards and regulations; namely NFPA 306, Control of Gas Hazards on Vessels, and OSHA and Coast Guard regulations.

NFPA 306, INDUSTRY CONSENSUS STANDARD

NFPA 306, as are all NFPA codes and standards, is unique because it represents the collective efforts of industry volunteers, who through the consensus standards-making process, develop the requirements for their industry. The Committee on Gas Hazards consists of industry representatives from the vessel owners, repairers, insurers, inspectors and regulators. Together, the committee establishes, with input from all interest groups, practical guidelines that ensure the safety of their industry. The federal government participates in the process and is then encouraged to utilize the voluntary standard in federal regulatory projects. NFPA 306 is one key to the maritime confined space safety program.

This standard applies to vessels carrying or burning as fuel flammable or combustible liquids. It also applies to vessels carrying or having carried flammable compressed gases, chemicals in bulk, or other products capable of creating a hazardous condition. This standard describes the conditions required before a space may be entered or work may be started on any vessel under construction, alteration, repair, or for shipbreaking. Work activities to which this standard applies include cold work, application or removal of protective coatings, and work involving riveting, welding, burning, or like fire-producing operations.

This standard applies to vessels while in the United States, its territories, and possessions, both within and outside of yards for ship construction, alteration, ship repair, or shipbreaking. The standard applies specifically to those spaces on vessels that are subject to concentrations of combustible, flammable, and toxic liquids, vapors, gases, and chemicals. This standard is also applicable to those spaces on vessels that may not contain sufficient oxygen to permit safe entry.

The purpose of this standard is to provide minimum requirements and conditions for use in determining that a space or area on a vessel is safe for entry or work. The standard accomplishes this through the application of "standard safety designations". These designations address the primary activities of worker entry and repair involving hot work and establish a framework within which the specific work activities may proceed safely. The designations include, **"SAFE FOR WORKERS", "SAFE FOR HOT WORK", "ENTER WITH RESTRICTIONS", AND "INERTED".** Each of the designations involves not only testing of the atmospheric conditions within the work areas, but also a physical inspection of the work area to determine if any potential exists for the conditions to change during the specified and approved work.

NFPA 306 requires that the Marine Chemist perform tests of the atmosphere and perform an internal inspection of the spaces. In addition, his determination of the condition of a confined space must include the following:

     o the three previous cargo loadings

     o the nature and extent of the repair work

     o starting time and duration of the work

o tests of cargo and vent lines at manifolds and accessible openings

o verification that valves are locked and tagged as necessary

o tests of cargo heating coils

NFPA 306 is incorporated by reference in the following OSHA and Coast Guard regulations:

OSHA 29 CFR 1915, Subpart B

USCG 46 CFR 35.01-1(c)(1), 71.60-1(c)(1), 91.50-1(c)(1)

The system of recognition, evaluation and control, links the NFPA Standard to the requirements found in the OSHA regulations. In each case, both NFPA 306 and the OSHA regulations identify the spaces in the marine construction, repair and shipbreaking industry, that are confined spaces and require testing and evaluation by the Marine Chemist and the OSHA designated "Shipyard Competent Person". NFPA 306 and 29 CFR 1915, Subpart B take a similar approach to identification of the specific confined space hazards which are likely to be encountered in the spaces identified above. These typically are listed as oxygen deficiency, flammability, and toxicity. These are the most common hazards associated with confined spaces, in general industry as well as in the marine industry. These hazards may occur as a result of the previous product (cargo or fuel), as a result of a specific work activity (welding, burning, sandblasting, paint spraying), or as a result of some reaction between residues or materials inside the confined spaces (such as hydrogen sulfide formation due to decomposition of organic matter).

The key to confined space safety, whether performing entry for inspection or cold work or performing hot work, is to evaluate the conditions within the confined space. This involves measurement of the hazards and comparison to the various guidelines. The guidelines used by the marine industry are:

o oxygen - normal air contains 20.8 % oxygen by volume, standards require at least 19.5 % oxygen by volume

o flammability - less than 10 % of the lower explosive limit (L.E.L.)

o toxicity - less than permissible concentrations, either the Threshold Limit Value or Permissible Exposure Level

Control of those hazards which have been recognized and evaluated presents the greatest difficulty for the Marine Chemist and industry together. Perhaps one reason for this is the normal tendency for workers to focus upon the immediate needs and conditions. All too often, when confined space accidents occur, the cause has been failure to maintain safe conditions throughout the duration of the prescribed work activity. Therefore, it is essential that industry and the Marine Chemist work together to promote conditions in the workplace that ensure worker safety initially as well as for the long haul. Control becomes the ingredient by which that objective can be attained.

Control is an action taken to eliminate or minimize a hazard that has been recognized and evaluated. Controls may be engineering, administrative or personal protective. The most common form of engineering control is ventilation. Ventilation is used during the initial gas-freeing phase when preparing a confined space for entry and work, and ventilation is also used during the entry and work to ensure that conditions do not change from those that were originally found when the entry and work were permitted.

Administrative control includes training, standard operating procedures or safe work practices, and permit systems that authorize work or entry. A safe work practice provides all the steps of the process that has been described in this paper: recognition or identification of confined spaces and hazards, procedures for evaluating and monitoring confined space hazards (determining the severity of the hazard), and methods for controlling known and unknown hazards in the workplace. The NFPA Standard and OSHA regulation provide the foundation for a very complete safe work practice. It forms the basis for developing training and educating workers.

The last method for control is personal protection. This should be considered as the last method of choice, since it means that the contaminant (hazard) has not been removed or eliminated from the workplace. Instead of making the space safe for the worker, the worker, by wearing some form of protective equipment (clothing or respiratory protection) has been made safe for the space. It is easy to see that if the protection fails, the worker is exposed.

It is essential to remember that part of control includes frequent monitoring and re-evaluating of the atmosphere or workplace. Just because a control measure (ventilation, protective clothing, respirator) has been implemented and is effective initially, does not mean that it will remain that way for the duration of the work activity, particularly if new contaminants are being generated as a result of the work. The key to working safely in confined spaces is continuous re-evaluation of the initial conditions. This includes the nature and extent of work, as well as the atmospheric contaminant level which requires periodic re-testing.

TRAINING

The importance of the industry and government cooperation has been stressed throughout this paper and credit for the program's longevity can be traced to this long-standing relationship. The joint commitment to safety within the industry prompted both parties to cooperate in the development of a training program for the OSHA designated "Shipyard Competent Person". With industry and government technical and financial support, NFPA coordinated and completed the development of a three-day, hands-on, intensive training course designed to provide workers with the ability to recognize, evaluate, and control confined space hazards that are present in the marine vessel construction and repair industry. The training has been designed to provide competent persons with an understanding of their responsibilities as found in the OSHA regulations and their relationship to the Marine Chemist. The training also provides hands-on skills in using and interpreting the results of the various testing instruments available for evaluating the atmospheric conditions within confined spaces.

OSHA, Coast Guard and industry contributed both financial and technical support to the course development. The project was completed in 1987; since then nearly one thousand students have completed the training at various port locations around the United States. The industry continues to underwrite a portion of the total costs for each course in an effort to contain costs to individual industry members. The course's success has prompted development of a follow-up course; presentation will begin in late 1990.

SUMMARY

Nearly seventy years ago the marine industry created a program in response to the immediate problem of large losses due to fire and explosion during vessel construction and repair. Today, this program thrives and serves as a model for general industry in the United States as OSHA prepares to deliver its final rule on confined space safety. The program succeeds because of the active participation of so many segments of the industry in OSHA and Coast Guard industry advisory committees; NFPA technical committees; the Marine Chemist Qualification Board; the Marine Field Service Advisory Committee; and activities sponsored by the various members. In order to understand the value of the industry's approach to confined space safety it is necessary to review the key program components:

- o OSHA Regulations
- o NFPA Consensus Standard
- o NFPA Certificated Marine Chemist
- o OSHA Shipyard Competent Person
- o Permit entry system - Marine Chemist Certificate
- o Engineering Control Practices - ventilation

The elements above comprise the basis for a technically sound and effective program for providing safety to all workers required to enter and work in confined spaces. When the elements of the program are in place, accidents do not occur. It is for that reason that the industry, in cooperation with the regulators, has placed such emphasis on training; to ensure that all workers know all the elements of the program and question any situation or condition which deviates from the requirements of the program.

The Marine Chemist, as the certified professional, becomes the focus of such a program. He provides training and guidance to those implementing the program, and functions within the specific requirements of the standard and regulations to perform required tasks. Because the problem in the industry has expanded beyond just fire and explosion, the Marine Chemist has continued to expand his training and expertise. Training for Marine Chemists includes heavy emphasis on industrial hygiene topics and broader safety and health issues. As a result, many Marine Chemists now possess dual certification as Marine Chemists and Industrial Hygienists or Certified Safety Professionals. Many find that their skills and knowledge are in demand beyond the strictly maritime workplace. They are able to transfer their expertise to problems involving fire and explosion, confined space safety, or atmospheric testing within other industrial settings.

The marine industry possesses what no other industry segment has - a professional engaged in day-to-day interaction with entry and work in confined spaces. Accidents result when individuals fail to recognize that confined space entry includes not only entry, but work and exit. The maritime program, with the two-tiered system for testing provided by the Marine Chemist and Shipyard Competent Person, is designed to ensure that workers are afforded conditions that are safe for <u>ENTRY, WORK AND EXIT.</u>

The task for the marine industry is to continue to foster the relationship that it has developed with the regulatory agencies; to promote wider understanding and acceptance within its industry of the requirements contained in the NFPA standard and OSHA and Coast Guard regulations; to further emphasize the importance of training; and to disseminate information concerning problems and changes in practices that may be of interest to all participants in the program.

The overall goal of this program remains the promotion and preservation of safety to life, limb and property within the maritime industry. NFPA, the Marine Chemist, and industry are prepared to meet future challenges to the program, together.

A REVIEW OF THE SAFE EVACUATION OF PERSONNEL FROM OFFSHORE INSTALLATIONS BY TOTALLY ENCLOSED MOTOR PROPELLED SURVIVAL CRAFT (TEMPSC)

C.WILSON

Offshore Marine Engineering,UK.

Time - Not on your side

In any emergency, there is virtually always one common factor - a shortage of time for people to reach a position of safety.

This is particularly the case in the need for the evacuation of an offshore installation.

The purpose of this paper is to review the evacuation by means of a totally enclosed motor propelled survival craft and how this evacuation can be as safe and as fast as possible.

It is the author's opinion that manufacturers of survival craft should do much more than meet existing legislation. We consider that legislation is only a starting point and should be the firm foundation on which to develop and introduce safer and more efficient evacuation systems. Offshore operators should also play an important role in developing systems that ensure that both able and injured personnel can be evacuated to a safe location in the shortest possible time.

Evacuation by by a survival craft has three key elements. Legislation tends to review these separately or not to address particular areas. The main time elements to be considered:-

☐ The actual embarkation of able and injured personnel

☐ The safe lowering and launching of the craft

☐ The time taken to reach a safe distance, considering adverse sea and wind states

The purpose of this paper is to introduce a major new safety feature for survival craft. It is not to discuss the merits of different type of launching system, these are discussed where applicable.

Detailed technical information has not been included with this paper, since the clear objective is to demonstrate how improved safety can be achieved to all levels, and not to just those involved with technical design.

A Short History of TEMPSC

Totally enclosed survival craft were developed to provide higher safety to the occupants of a lifeboat. This has evolved from pure protection from the elements through to the ability to withstand fire and self-right in the flooded condition.

A number of craft were simply developed from existing open lifeboat hull forms by the addition of a canopy. Whilst this proved to be to effective, obviously, there were a number of limitations, especially in access for embarkation, occupancy and the provision of the latest safety features.

Currently, there are two systems available, freefall and conventional. This paper will concentrate on the conventional form, since in the writer's opinion, the developments reviewed in this paper provide the most enhanced safety available.

Both conventional and freefall TEMPSC have progressed in design and ability over the past years and incorporate many lessons learnt the 'hard way'.

In July 1986, the latest International Maritime Organisation (IMO) amendments came into operation. These have further enhanced boat design and improved both performance and safety. At present, there are a number of conventional craft that carry certifications to the latest IMO requirements. To date, no freefall TEMPSC has obtained DoT approval for use on fixed platforms in UK waters, but this is expected to change in 1991.

The Latest SOLAS Rules and Testing

Several important changes to all lifeboat designs were made in the IMO 1983 amendments - especially for totally enclosed versions. Here the rule changes meant that many existing designs could not be used.

One of the most important and fundamental rule changes concerned the self-righting ability when damaged and open to the sea. The new regulation requires that a lifeboat should be capable of righting itself when capsized and should provide an above water escape for the occupants. This compares with the previous regulation which required self-righting in the intact and undamaged condition. Although a simple sounding concept, this requirement changed significantly the shape of lifeboats.

Another new addition was for the occupants to be protected from dangerous accelerations arising from impacts during launch by four point harnesses and head restraints. This involves the prototype of each new design to be subjected to a side impact of 3.5 m/s, which replicates a collision with a ship's side during launch.

Fire protection of lifeboats is not new, but the revised regulations require a duration of 10 minutes for the on board air supply. Both the occupants and engine require to be supplied from a bank of air bottles which are regulated to provide a positive air pressure within the boat. This prevents smoke and gas being drawn into the craft whilst the lifeboat proceeds to a place of safety.

The fire protection system includes an engine driven pump that draws sea water and sprays it over the hull to prevent fire damage to the craft.

Important changes were made to the operation of the davit release system. The new regulation requires an on-load release capability. This feature is of great benefit when a lifeboat is launched from a ship that is making way.

All TEMPSC located on fixed offshore installations within the UK continental shelf require to be approved by the Department of Transport (Marine Directorate) acting as agents for the Department of Energy.

The Development of the TEMPSC

The impending changes in the rules on lifeboat design led manufacturers to formulate a completely new design.

For the author's company, the hull design was in fact to be dual purpose and was designed from the inside out, i.e. with the safety of the occupants first and the hull shape second. This compares to many existing craft which were based on open lifeboats with a canopy and little thought to the safety of the personnel.

Engineering experience in the design and construction of systems for the offshore sector has led to a firm belief that standards and regulations are only a starting point in design. Only development of designs that show what is possible with given technology brings about pressures for updates and changes to existing regulations.

One new design concept was to produce a dual purpose hull, which could be used for a range of passenger lifeboats and also as the basis for evacuating pressurised divers from decompression chambers on diving support vessels.

The version for divers is known as an SPHL (Self Propelled Hyperbaric Lifeboat) and incorporates a large chamber for the divers and all associated life support facilities including the special breathing gases. The lifeboat is crewed by trained operators who are responsible for launching and reaching a place of safety. The lifeboat crew occupies an area behind the pressure vessel where all the equipment to propel and maintain life support functions is located

Hull

A glass fibre laminate design can be hand-laid mat construction or the more common chopper-gun method. The hand-laid method is considered by the writer to be superior where a service life of 20 years in arduous conditions is expected.

Exceeding The Regulations

In a number of areas, the design by various manufacturers exceeds the current SOLAS regulations.

One example of this is the provision of an above water escape after a flooded capsize. This implies that the boat may have an extreme list or trim and be so swamped that the occupants have no choice but to abandon the lifeboat.

If you consider that SPHL version, the divers are located in a pressurised chamber and have no choice but to remain in that chamber until it can be connected to a decompression system. To this end, the design of the buoyancy system must be designed so that the doors on both sides are clear of the water after a flooded capsize. This would enable the crew to bale out the boat and continue to provide attention to the divers.

This is one example where the concept from one type has been transferred the other. The concept has been maintained on the range of passenger lifeboats. All designs have all doors clear of the water when fully flooded and withstand the weight of the people standing on the sills without excessive heel. This permits embarkation to a rescue helicopter without creating the danger of further capsize.

Another significant development is the method of safely evacuating the survival craft occupants to a helicopter. Here, a design has been produced which is not only available as standard by the manufactuerer, but can also be retro-fitted to a number of other types of craft. To gain the maximum degree of benefit and safety from such a system, it is essential that the survival craft has a clear exterior with no major protrusions which could snag the cable lowered from the helicopter. The design that has been developed on new craft provides for a convenient interior platform, a large opening top hatch which accommodates up to three persons and suitable access ladders and hand points. It is located in the forward part of the craft and allows the coxswain full view of the rescue operation. This design has been subjected to extensive testing and benefits from previous experience in helicopter rescue systems.

Interior

The interior of a TEMPSC must be designed to withstand the treatment expected in service. All interior fittings, such as seat frames and equipment lockers, should bolted to foundation plates bonded into the glass-fibre structure.

Great emphasis must be placed on maintaining an interior style that is common to the full range of lifeboat designs of various sizes. This has an important advantage in crew training as the layout, controls and instruments are similar, regardless of the craft size. This reduces the requirements for operators to train crews on a specific size of lifeboat.

The interiors of all craft should be bright and airy and laid out to make the occupants as at ease as possible. For example, bright conditions should be achieved by the cabin lights at night or the translucent hatches by day.

Crew confidence must not be overlooked in an emergency. Dark and cramped interiors cannot be beneficial to the well being of those whose lives are at risk and who may be frightened, injured or suffering from shock. Confinement in a lifeboat should be as comfortable as possible and must not exacerbate the condition of the occupants.

Factors to be considered in Evacuation

In order to obtain the maximum safety achievable, the following factors should be considered when selecting and installing TEMPSC.

Craft Location

An obvious consideration is not just the capacity of the craft to meet the regulations, but also the distribution of the craft. The first consideration must be to make the TEMPSC readily accessible for the crew, throughout the whole installation and especially from Safe Haven Refuge.

Access

Having reached the TEMPSC, embarkation must be achieved in the minimum length of time for both able, injured or crew impaired by shock. Legislation in the UK sector requires for a minimum length of time for embarkation. However, embarkation times vary considerably for different manufactured craft. Time variations are approximately 1-45 minutes to OVER 3-30 minutes for a 50 man TEMPSC.

The actual internal layout of the craft and the entrance doors should be designed to allow both able crew and stretchers to be secured in the minimum length of time.

Lowering

This paper concentrates on conventionally launched TEMPSC, the advantages of which are considered in another part of this paper. Launching systems, including davits have been refined over a number of years, but sometimes are not considered as part of the evacuation system at the design stage.

Craft Orientation

The actual method of deployment of a TEMPSC plays a vital role in enhancing clearance from the installation. If a craft is positioned parallel to the installation, then undoubtably there are major problems under bad sea and wind conditions in successfully clearing the installation. To achieve the maximum safety, TEMPSC should be mounted perpendicular to the platform, i.e. pointing outwards and away.

Performance of the TEMPSC

The actual ability of a TEMPSC to clear an installation in the minimum time does not form part of any current regulations. The only performance requirement is for a TEMPSC to achieve a speed of 6 knots in calm water.

Clearly, there have been considerable improvements in overall safety in all the above areas - apart from performance. A key element in ensuring the safety of the occupants must be for the TEMPSC to achieve high acceleration so that it can reach a place of safety in the minimum length of time. This vital consideration is reviewed in detail in the next sections.

A Summary of the advantages of conventionally launched TEMPSC

Whilst it is not the purpose of this paper to compare conventional and free fall launch systems, it is worth considering some of the advantages and disadvantages of conventionally launched TEMPSC.

The principal advantages of TEMPSC are :-

- There are a number of manufacturers with craft approved by the UK authorities and these craft cover a comprehensive range of capacities.

- There is considerable experience and technical information available for high launch situations.

- There is operating experience of launching in severe wind and sea states.

- Compared to a freefall launch from a fixed installation, there is a very low risk of collision with debris, vessels, rescue craft, other lifeboats and even personnel in the water.

- Injured personnel are not placed in an unacceptable risk during the launch.

- Embarkation time is reduced, thus decreasing the risk and exposure factors.

- The layout of the access hatches on conventional craft provide easier access for the recovery of persons from the sea. In general, conventional craft hatches are not restricted to stern doors only.

- A heli-hatch system is available and can be retro-fitted to conventionally launched craft. The heli-hatch system provides for the rescue of the occupants of the lifeboat by helicopter and has been tested extensively.

- The training requirements of conventionally launched craft are much smaller and there is a reduced need for on-going and in-depth training.

- They provide the minimum weight and minimum cost options.

There are a number of disadvantages with conventionally launched craft. Some of these have been discussed previously, but are worth re-stating.

- The position of the craft in relation to the platform can affect the evacuation. If parallel to the installation, then there could be major problems under bad sea and wind conditions.

- Embarkation times vary considerably from one type of craft to another, with some craft having unacceptably high times.

- The layout of access doors on some craft do not facilitate the loading and removal of injured personnel who are confined to a stretcher.

- The latest generation of both conventional and freefall craft tend to have inherent acceleration and manoeuvrability problems.

With the exception of the last item, these disadvantages have been overcome by good design practice. The reasons for this very important shortcoming is discussed in more detail in the following sections.

The Performance of Current Craft

It is generally accepted that all of the latest generation of TEMPSC, both freefall and conventional, have inherent acceleration and manoeuvrability problems. Whilst their general speed performance may be acceptable, this maximum speed can take a considerable time to reach - i.e. poor acceleration. To achieve a safe distance after launch, much improved acceleration times are required. The performance of current craft are mediocre in other areas, including steering astern, particularly when turning against the propeller torque, and across the wind and/or sea.

The reasons for the indifferent manoeuvrability and poor performance are fairly obvious

- ❑ Low L/B ratio, i.e. a traditional short, fat hull form
- ❑ A large canopy windage
- ❑ Relatively low installed power to meet a specified speed
- ❑ Poor water flow into the propeller disc.

These characteristics, which all modern TEMPSC share to some extent, are effectively the consequence of the regulations which these craft have to satisfy. They are also the result of the demands of the marine and offshore industry for optimum seating capacity within a given boat size.

In the shipping industry, where TEMPSC originate, manoeuvrability is not such a critical requirement, as after launch the craft may run down the side of the ship until it has sufficient way to steer clear.

Clearly, this is not the case offshore, as it is vital that the craft should be able to manoeuvre clear of the jacket structure from a 'standing start' and AGAINST the wind and sea if necessary.

Even when stowed 'outward facing', conventional craft may experience some of the above difficulties in successfully clearing the platform in adverse conditions.

The Twin Engine TEMPSC - The Way Ahead

A research and development programme has been undertaken by one manufacturer on how conventional craft can be improved, with particular emphasis on the reduction in time to reach a point of safety from the installation.

The results of this programme have been the development of the TEMPSC Twin, a twin engine variant of the current Phoenix design which is fully proven and in service with a number of operators.

Engines

As an example, the 60 man version utilises two 40 HP marine diesels. The units selected are in fact developments in a new range of engines approved by the Department of Transport for SOLAS lifeboat use.

An important element of the design requirements was that there should be a minimum of changes made to the arrangement of the craft and that the maximum number of existing components should be re-used, unchanged or with minor modifications. Other important considerations were that the capacity of the craft should not be adversely affected and that access and embarkation times would remain the same.

The two engines are installed on a common sub-frame and enclosed in fire retardant GRP/foam sandwich housing which provides both heat and sound insulation. Hinged covers, similar to the existing craft, are fitted for access.

In between the two engines, the space is used to house the deluge pump. sea suction, keel cooler valves, the hydrostatic release unit and other smaller equipment. This has the advantage of concentrating all machinery in one accessible area.

Propellers and Steering

The engines drive handed outward-turning propellers fitted in fixed nozzles supported on 'A' brackets. Steering is via a plate rudder fitted in place of the steerable nozzle currently in use. This nozzle was fitted to production craft to improve propulsion efficiency and was made steerable to improve manoeuvrability astern. With twin screws, this nozzle is not required, since the craft can be steered by means of differential throttle, gearbox and rudder.

Layout

With the design adopted, there is no significant changes forward of the engine housing or at the extreme aft of the craft. By careful design, only one part of the TEMPSC is changed and all other items such as seating, hatches, fuel tanks, air system storage tanks and similar items remain the same.

The net loss of seating due to the inclusion of the second engine is only a total of three. This number has been achieved by the relocation of certain items to the central machinery housing.

Benefits

The twin engine craft has all the proven benefits of an existing range including space headroom and in-built facilities for stretcher cases without the loss of seating capacity. The addition of the second engine greatly assists in overcoming the concerns and problems safely evacuating an offshore installation during adverse weather conditions. The Phoenix Twin provides the occupants with :-

- ❑ Greatly increased acceleration from a standing start
- ❑ 100% machinery redundancy
- ❑ Greatly improved steering and manoeuvrability

SUMMARY

An evacuation to sea during an emergency is both demanding and potentially very dangerous, especially in severe weather conditions.

It is essential that full consideration is given to the personnel to reach a safe location away from the installation in the minimum length of time. The key elements in achieving this are

- ❑ The location of the craft
- ❑ Access to the Craft
- ❑ Orientation and operation of the Launching System
- ❑ The actual performance of the craft when launched in clearing the installation.

With the advances made in the design and construction of conventional TEMPSC (including the latest IMO requirements, the following improvements have been achieved.

- ❑ Improved seating arrangements that provide a safer environment for the occupants
- ❑ More internal space for the occupants
- ❑ Four point harnesses and head restraints fitted for the personnel
- ❑ More rapid boarding has been achieved on certain range of craft due to improved access systems.
- ❑ Stretcher carrying facilities include, with no loss of seating capacity on some manufacturers range.
- ❑ Inherent self-righting capability in the flooded mode.
- ❑ On load release equipment allowing easier release in adverse weather conditions
- ❑ Maintenance and reliability have tended to improve due to improved design and equipment
- ❑ The heli-hatch rescue system is available and can be retro-fitted to many existing craft.

The twin engine craft developed by the author's company provides further safety for the occupants by :-

- ❑ Increased engine power
- ❑ 100% machinery redundancy (engines, deluge system etc)
- ❑ Improved acceleration from the 'standing start'
- ❑ Improved sea-keeping characteristics
- ❑ Improved manoeuvrability

Emergency Training Inside Industry

by George Carrol & Stewart Kidd

INTRODUCTION

The spate of disasters that have afflicted many sectors of industry and commerce over the past 5 years has highlighted the need for government, business and commerce to be prepared for the worst. The range of possible corporate disasters, which can easily destroy or severely damage even the largest of organizations has brought about a new interest in contingency planning.

The concept of preparing for the worst has been accepted by many companies but the extent to which it has been fully adopted, with all the resource implications of time, cost and training is difficult to assess. In fact many companies admit that they do not feel that they are susceptible to a major corporate crisis. The main reasons for their complacency are that they consider the sheer size of their organizations would overcome any problems and their safety standards are, in any case, considered very high.

In the United Kingdom, county and district councils (or their equivalent) have legal and social responsibilities to provide a degree of protection for the population. These responsibilities are in summary:

- ❏ to co-ordinate the response to a disaster

- ❏ to provide social and other services

- ❏ to organize restitution and recovery measures

The legal, social and moral obligations which all organizations have towards the local population are strengthened by legislation at certain industrial sites - *The Control Of Industrial Major*

Accident Hazards Regulations 1984 (CIMAH)[1]. They have a duty, in this respect, to make on-site emergency plans to deal with the effects of an accident. Local authorities are also required to make off-site plans at sites covered by the CIMAH regulations for accidents which are likely to affect the surrounding areas.

The organization and resources which are necessary for contingency planning have much in common. Moreover, major emergencies invariably occur without warning, and an organization prepared to meet any one of the possible contingencies is well placed to meet all of them. General purpose plans are therefore prepared and can be used to meet a major emergency of any kind in the county. This type of planning is known as the all hazards approach.

THE PROBLEM.

Current health and safety legislation[2] places a legal responsibility on the company on behalf of the workforce and public, but good contingency planning requires a wider perspective of the problems facing the whole of society in the aftermath of a corporate crisis. That such incidents can impact on the environment and community is well documented, particularly in the light of incidents at Chernobyl, Sandoz and the 'Exxon Valdez' spillage. What about the other longer term impact of a major incident? Flixborough amply demonstrates what happens when a major incident destroys a site. The products made at the plant had to be made elsewhere and, when the site was rebuilt and re-commissioned demand no longer existed because the competition had taken over. The whole social structure of the surrounding area was affected by the closure of what had been it's main employer. The scenario for which planning should be considered include:-

> Extremes of weather
> Loss of essential services
> Transportation accidents
> Major industrial accidents
> Corporate Crisis e.g.
>> *product contamination - accidental or deliberate*
>> *failures of sub-contractors*
>> *investment & financial difficulties*
>> *computer related eg failure, virus, crime*
> War

Under what circumstances is it appropriate to write specific plans or develop a wider approach and adopt an 'all encompassing' approach? As different circumstances, such as geography, resources, processes etc. affect the planning process, is there a case for specific plans? Producing plans which are narrowly based to fit detailed and specific scenarios is, by definition a limiting factor in a corporate response scheme.

It is clear from the foregoing that managers and senior engineers involved in any industry must be equipped with the necessary knowledge and expertise to be able to undertake an assessment of the disaster potential of their industry.

While current engineering degree courses do not, as far as the authors are aware, include such matters, there are a number of institutions offering short degree options courses and seminars on a range of matters related to risk.

Professional institutions would appear to be able to play a prominent role here. It is no unreasonable for society to be able to expect that a professional engineer who is deemed to be competent to be able to design a chemical plant should also be capable of recognition, analysis and assessment of the risks which his plant might pose.

It is suggested therefore that the techniques of risk management should be incorporated into al professional engineering training.

THE KNOWLEDGE TO COMBAT THE POTENTIAL PROBLEM.

Response to any major disaster will inevitably overwhelm the resources of even the largest company and where the consequences of a fire and explosion extends beyond site boundaries, there will be a considerable range and variety of interests involved. Some knowledge in the areas mentioned in the following paragraphs is therefore necessary,

Organization and roles

An important objective, particularly in the context of local authorities, should be the organization and roles of the police, fire and ambulance services. There are many differences in the way the emergency services operate. For example, different police forces have different operational procedures, as do the fire services - and their equipment can also vary. However, some common ground can be covered and local differences dealt with individually.

The roles of local authority departments, the public utilities and government departments with an emergency planning responsibility such as the Ministry of Agriculture, Fisheries and Food must also be understood. Also included should be the role of the voluntary aid societies such as St. John Ambulance, British Red Cross Society and the Women's Royal Voluntary Service.

Legislation

Detailed knowledge and understanding of national and, increasingly relevant, European Community legislation such as Civil Defence regulations, Control of Industrial Major Accident Hazards regulations and Transportation of Dangerous Cargoes is essential.

Role of the Military in Civil Emergencies.

This area should cover the role and range of military aid available, the legal implications of using military resources and the cost of such aid. For those without a service background, an introduction to the organization and structure of the three armed services is essential.

Economic and financial.

An investigation into the way authorities are funded and how they apply for and account for grants from central government. Financial management is an important element for both the local authority and the industrial planner.

Political and environmental

Politics play a major role in the emergency planning field, at local and national level and will affect the industrial planner as well as the local authority planner. The political aspirations of 'nuclear free' authorities affect not only the local authority planner but those involved with the use and movement of radioactive materials, particularly in relation to electricity generation. There are international implications, especially European issues, and the need for co-ordinated international action; for example, ferries, road transport and the Channel tunnel project.

Just as environmental issues have become more closely involved with politics so must planning take into account any impact on the environment. The 'Sandoz' incident at Basle in November 1986 demonstrate this admirably - particularly with the benefit of hindsight. Over 30,000 litres of water per minute were pumped onto a fire and the only place the now-contaminated water could drain into was the Rhine. Compounding the problem, contaminated water was pumped back onto the fire resulting in the production of mercaptans, and other toxic compounds. The twice-contaminated water 'killed' the river for most of its length and resulted in water abstraction being terminated for some time.

In a second incident, a fire lasting 11 days at a cold store warehouse in Hamburg, Germany, vast amounts of butter and fat emulsified with the run-off water with resultant damage to canals and lakes killing substantial numbers of birds and fish. The sewage system and many cellars were also filled with fat.

The fire and explosion at the Chernobyl nuclear power station also demonstrates the far-reaching consequences of disasters - and not only to the nuclear power industry.

The Confederation of Fire Protection Associations - Europe in a report stated[3]:-

> "The increasing threat coupled with this growing public concern have brought new responsibilities to bear on owners and managers of companies and to the fire officers who have to deal with emergencies. Not only must fires be contained and extinguished, but thought has to be given to the consequences of the fire and fire fighting for the neighbourhood and the environmental risks they hold for the people, soil, water and air."

This quote is equally applicable to those with an emergency planning role.

Civil Defence.

The widening of the emergency planning field cannot ignore the responsibility for civil defence which every government has. Just as industrial targets are considered important to the economics of war so are they relevant to the industrial emergency planner. The threat terrorism poses can be compared to war and should not be overlooked.

Information Technology.

The development of faster computers with larger memories allows vast amounts of data to be stored and recovered with speed. Geographical information can be linked to databases allowing the effects of clouds of toxic products to be plotted and displayed graphically. These developments are already providing excellent tools for the emergency planner. To make better use of these tools and to help develop future systems an awareness and understanding of this technology is increasingly important.

EXERCISE WRITING.

The only way plans can be tested is by exercising them. Increased awareness of and confidence in the plans, and the planner, are an important consequence of the exercise, therefore the ability to organize and run exercises is critical. To run a successful exercise three basic elements are required namely:-

❑ research

❑ preparation

❑ management support

Research

Time is often so crucial in emergency response that much detailed planning is necessary to ensure that resources are brought to the scene in a timely and logical fashion. It is surprising how often a delay can occur if such mundane matters as-call out procedures for key decision makers are overlooked, there will always be an element of the unforeseen but careful and detailed research into all aspects of organization should include:-

Management
Business Plan
Departments and inter-links
Processes
By-products
Key figures and Job descriptions
Call out times for emergency services
Legislation

Investigation into many major accidents has identified the need to appoint an emergency management team from the outset. The research phase will help identify those best suited and in particular a *LEADER*.

Preparation.

An exercise may be treated in much the same way as a project is in industry. and, like most projects, it is important to define or identify the aim of the exercise. The purpose of exercises may be to:-

     a.   test

     b.   teach

     c.   identify training needs

     d.   stimulate interest

     e.   combination of two or more of above

The project may be managed by an individual or a team

Once the aim is established, objectives and their resultant parameters should be agreed and written in the exercise instructions. Omission of the aim and objectives will, invariably, result in a project that is impossible to evaluate or assess.

The next stage is to plan and write the exercise and to decide whether to exercise all or only part of the organization. A scenario is usually written to encourage some form of realism and to help in the decision making process. This stage should be treated with caution, however, and based within the bounds of possibility and within the stated aim and objectives. By describing an improbable scene the credibility of the exercise becomes questionable. The scenario may be added to, in real time, as the exercise unfolds but this must be a planned input to gain maximum benefit.

Pre-exercise administration is an important stage as success or failure is dependant on sound preparation. All concerned should be aware of what is expected and all the required equipment be identified and made available.

Most exercises require several persons to oversee the action or to input new elements following the initial scenario, they are often called the directing staff. The directing staff may also act as safety officers, ensuring that what begins as an exercise does not result in a real emergency or injury. They should be appointed, preferably at an early stage, and their role and tasks rehearsed, or at least briefed - including their safety role.

A full and comprehensive briefing before the exercise for all concerned, including observers, is paramount to its success. On some occasions, particularly in testing exercises, it may not be advisable to have all present only a representative who will supervise the area of his responsibility. Directing staff may often need supervision or advice and guidance during the course of the exercise so consider some form of separate communications for and with them.

Immediately at the end of the exercise a de-brief to highlight the lessons learnt, both positive and negative, while still fresh in everyone's mind is advisable. Encourage feedback from all including observers, directing staff and, importantly, those being exercised and make written notes of the comments.

Post exercise administration is necessary to ensure that any interruption to the operation of the organization is kept to a minimum. All equipment used is made serviceable and returned, with the scene of the exercise restored to it's original state.

Finally evaluate the event and use feedback from participants to identify aims or objectives for next exercise.

Management Support.

Writing and planning the exercise is probably the easy part, convincing management on the value of the training is an entirely different matter. Without the support of senior management the project can only achieve a limited part of its aim and the lessons learnt being treated with scepticism. A sound business case must be made for the project based on legislative, financial and resource arguments. Decisions on the scale and scope of the scheme when presenting the case for this form of training must be clearly presented.

One large, all encompassing exercise is expensive in time, resources and finances and will need a great deal of training to prepare for it, so it may be wise to consider a series of smaller exercises to develop the different parts of the organization. It would also be prudent to include members of senior management in all aspects of the project from planning to implementation and even through to evaluation.

IMPLICATIONS OF FAILING TO PROVIDE ADEQUATE TRAINING.

Legislation.

Everyone owes those who may be affected by their acts or omissions a general duty of care. A breach of this duty of care not to harm someone may constitute negligence - and an action may be brought by the victim of such negligence. The duty of care extends beyond action and includes advice. The extent of the duty is measured by the courts and as a consequence is difficult to define. Legal obligations may not be absolute but are confined to taking reasonable care to ensure that others may not not be exposed to foreseeable harm or suffer predictable losses. Training falls into the category of taking reasonable care.

The present legislative regime is based largely upon the the proposals of the "Robens Report" which resulted in the the enactment of the principle statute - *"The health and Safety at Work Act 1974"*. In addition to the prevailing Acts of Parliament there are also many Orders and Regulations in force. Part 1 section 2 *"General Duties"* paragraph (2)(c) states:-

> *"the provision of such information, instruction, training and supervision as is necessary to ensure, so far as is reasonably practicable, the health and safety at work of his employees"*

The health and safety at work act, however, must not be read in isolation. It must be considered in the light of earlier and subsequent legislation and by the attitude of the judiciary. Indeed, a former Chief Executive of the Health and Safety Executive commenting on the HSE's role:-

> *"... in view of its resources should focus on behaviour, training and supervision at the workplace".*

Another legal area is that of the contracts of service which constitutes a true contract of employment. The terms binding the parties to contracts, which could include the contract to provide adequate training are to be found in:-

Written and verbal agreements (expressed and/or implied).

Rule books

Custom and practice

Terms and conditions of service

Statutory provisions

Collective agreements

etc.

Financial

The authors are not aware of any insurers that are currently offering premium discounts to offset the cost of training but the result of failing to comply with legislation could result in the negation of an insurance policy, which itself might be a statutory requirement to do business.

SUMMARY.

Corporate and financial planners are an accepted part of any business strategy and contingency planning and crisis management should be no exception. The large percentage of businesses which cease to trade following a major incident, whether natural or man made, should serve as a warning to the remainder. Planning to limit any interruption to business is important and particularly so in a highly competitive industry - regardless of the size of the organization. Business as usual may not always be possible but well made plans should restore some form of normality with the least delay.

It has been demonstrated that companies with an awareness of the hazards of their business and contingency plans to deal with those hazards, are well placed to meet most situations. They are also able to show that they have taken all reasonable precautions to reduce the likelihood of the incident happening in the first place.

All exercises must have a clear aim with logical, attainable objectives. Only then can training and exercising highlight the shortcomings of an organization, but that training must be relevant, challenging and effective.

The recent publication of the report into the Piper Alpha incident should serve as a reminder to us all of the dangers of complacency.

References.

1. **Control of Industrial Major Accidents Regulations 1984**
2. **Health and Safety at Work Act 1974**
3. **Fire and Its Environmental Impact**
   CFPA Europe: 1990
   The Loss Prevention Council

THE APPLICATION OF FORMAL SAFETY ASSESSMENT TO AN

EXISTING OFFSHORE INSTALLATION

M P Broadribb
Central Safety Engineering Superintendent
BP Exploration

1. INTRODUCTION

In the wake of a disaster such as Piper Alpha there is often a call for engineering solutions to prevent a recurrence. Yet, experience shows that inherently dangerous activities can be safe if well managed, yet common and everyday activities can be disastrous if handled negligently. In most tragedies, people and procedures, not just the hardware, play a large part.

However, the offshore industry has always recognised the need to design into facilities a high level of inherent safety and has traditionally relied upon sound and proven standards and codes, audits, management supervision and control, etc. in a very straightforward, logical and structured way to do this.

Recognising the special hazards in the North Sea environment and in particular the potential for extreme events such as ship collisions or fires and explosions involving large numbers of people, many operators over the years have incorporated more specialised engineering tools and techniques.

Even before the Piper Alpha incident, various individuals within the industry were concerned that reliance on good engineering practice, the application of approved standards and the certification and inspection regimes did not of themselves comprehensively identify and highlight the hazards and sequences of events that could lead to a major accident. Whilst use of more specialised engineering tools and assessment techniques is certainly appropriate, the Department of Energy (DEn) have proposed Formal Safety Assessment (FSA) in order to address overall safety problems.

In October 1989 the DEn Safety Directorate issued a Discussion Document "Formal Safety Assessments of Offshore Installations" (Ref. 1), describing a process of continuous safety assessment, addressing the installation hardware, the human factors, operating practices and procedures and the safety management regime. Subsequently during Part 2 of the Cullen Inquiry much of the evidence supported the concept of FSA, and, as is now widely known, Lord Cullen emphatically endorsed this approach in his recent report (Ref. 2) by recommending the preparation of an offshore Safety Case.

The United Kingdom Offshore Operators Association (UKOOA) has strongly supported the concept of pulling together all these relevant engineering and management tools in a formalised manner and that this should apply to all future and existing installations. Some operators are already committed to this concept and are making progress in implementing it. It is an approach which provides not only a systematic and fully documented analysis, specifically from a safety point of view, of the original engineering design, it also serves as a living process to assess safety right through the construction, operations and decommissioning phases.

2. **FORMAL SAFETY ASSESSMENT**

To date, of course, there are no specific regulations requiring and defining the preparation of offshore Safety Cases. Indeed a formal Consultative Document containing draft regulations will not be published for industry comment until perhaps late 1991. The final regulations are therefore unlikely to be enacted prior to mid-1992.

The interpretation of the terms 'Formal Safety Assessment' and 'Safety Case' is therefore a matter of individual preference, vested interest and sometimes heated discussion between various parties! However the consensus that emerged within UKOOA, and was subsequently endorsed by Lord Cullen in his report, was that the FSA is analogous to the Safety Report (nee Safety Case) required by the onshore CIMAH Regulations. As such the objectives of the offshore Safety Case should be similar in nature, viz:

i)   to give an account of the arrangements for safe operation of the installation, for control of serious deviations that could lead to a major accident and for emergency procedures at the installation;

ii) to identify the type, relative likelihood and consequence of major accidents that might occur, and

iii) to demonstrate that the operator has identified the major hazard potential of his activities and has provided appropriate controls.

The author would like to suggest that the term 'Formal Safety Assessment' can be used to describe the 'process' whereby the above objectives are met, and that the 'Safety Case' is the 'deliverable' at the end of the process.

In keeping with the concept of the FSA being a living process encompassing all stages of the life of a development, the Safety Case should be reviewed and updated at key stages. The key stages envisaged are:

(1) Concept

(2) Detailed Design (including Construction and Operation)

(3) Review of (2) during Operation (at regular intervals and in event of significant changes)

3.   CONTENT OF THE SAFETY CASE

A Safety Case is a demonstration that the activity can be carried out safely; it includes a description of the major accident hazards that could arise from the Operator's activities and the controls that are exercised to prevent them or to limit their consequences. As such the content of the Safety Case will vary with the different stages highlighted above, but will generally follow the subject matter in Table 1.

3.1   Concept Stage

At the concept stage the Safety Case will provide a broad
overview of the engineering determining the eventual layout and
configuration, and the management philosophy to be adopted.
All design options considered should be discussed, and the
reasons identified for the final concept selection.

3.2   Detailed Design

The next stage of the Safety Case at detailed design is
essentially the core document upon which subsequent reviews are
based. At this stage the Operator will have an in-depth
understanding of the mode of construction, operation,
maintenance and inspection. Therefore these aspects should be
fully addressed within this stage.

This core Safety Case is essentially an abstract of relevant
information about the major hazard aspects of the Operator's
activities drawn from a much more extensive body of
information. This body of information will include plant
design specifications, operating documents, maintenance
procedures and information derived from the Operator's
examination of his major hazard potential. In practice this
means that the Safety Case submitted will be a summary of work
undertaken by the Operator. It need not contain the detailed
documentary evidence which supports the conclusions reached but
it should include sufficient detail of them to enable an
independent assessor to judge whether the conclusions are
sound. It should also contain precise references to where the
supporting documentation can be consulted if necessary.

The information will fall into two broad categories: first,
factual information about the installation, its activities and
surroundings, and second, reasoned arguments and judgements
about the nature, likelihood and scale of potential major
accidents which may occur at the installation and the means to
prevent these hazards being realised or to adequately mitigate
their consequences.

The Safety Case should provide adequate justification for its conclusions, either by setting out the sources of the evidence for a particular argument, or by recording the principal assumptions in sufficient detail to enable them to be challenged if it emerges that they are critical to the conclusions.

3.3 Reviews

The reviews during operation should follow the same principles as outlined in 3.2 above. These reviews will concentrate upon any substantial differences, such as engineering modifications, procedural and management changes, simultaneous operations (drilling/production/construction) and abandonment.

4. MAJOR INCIDENTS

The discussion within the Safety Case should focus upon those events with major accident potential. The site specific nature and complexity of different installations will influence the potential major accidents and the measures taken to prevent, control and mitigate their consequences.

The following is a suggested check list of potential causes/initiating events of major accident hazards that all FSAs should consider:

- Blow out
- Riser/pipeline failure
- Process containment failure (including procedural failure)
- Helifuel release
- Turbine/compressor failure

Fire, explosion, pollution and toxic emissions should be considered for these hazards.

- Ship/boat collision
- Severe weather
- Dropped objects/lifting failures
- Helicopter crash
- Flotel/tender assisted drilling (TAD)/multi purpose support vessel (MSV) collision (eg. DP failure)
- Foundation failure
- Structural failure (eg. fatigue or corrosion)
- Drawdown (concrete structures only)

These are generally structural matters, although they could escalate to include fire, explosion etc. as above.

- Stability
- Mooring

These are only applicable to floating installations.

Appropriate combinations of relevant major accident hazards should be considered, to ensure that the potential for hazard escalation has been identified.

Some events do not in themselves directly result in loss of life/ installation damage. However, they can produce added complication and/or play an important role in the escalation of an incident. Examples of such events include:

- Communication failure
- Power supply failure
- Process control failure
- Diving
- Combined operations.

The effects of these hazards should be considered to evaluate their incremental contribution to the overall damage/loss levels.

5. SAFETY MANAGEMENT SYSTEM

An important part of the Safety Case will be a description of the Operator's Safety Management System (SMS). Indeed Lord Cullen drew particular attention to the requirement for a sound SMS in his report (Ref. 2). The term 'SMS' is probably a misnomer, as in reality the requirement is to demonstrate that 'the Management System' adequately addresses safety issues at all times.

In terms of the Safety Case, the SMS must ensure the identification and assessment of hazards throughout all offshore activities at all stages of development. Furthermore that all reasonably practicable measures are taken to prevent, control and mitigate those hazards. Finally the SMS must ensure that the information in the Safety Case is factually correct, and how the Operator will continue to observe the critical safety practices and features described.

6. SUBMISSION

An essential objective of an FSA is for the Operator to demonstrate to an outside authority that he has identified the potential hazards of his installations and has taken adequate measures to control the hazards. The role of the outside authority in examining the Safety Case should be to assess whether the Operator has fulfilled his duties to identify and prevent major accidents. It is therefore essential that the outside authority is competent in assessing both the engineering and management control aspects.

Due to the integrated nature of the Safety Case, there should be a single body responsible for the overall assessment. Lord Cullen has recommended that the Health and Safety Executive (HSE) assume regulatory control of offshore safety. The author is confident, given his experience of their approach to onshore operations, that the HSE will competently fulfill the above requirements.

7. **APPLICATION TO AN EXISTING INSTALLATION**

The range of installations to be addressed retrospectively will vary from those which have been in operation for many years to those which are currently under design/construction/commissioning. Even the development of installations currently under design/construction/commissioning may not have followed a strict FSA approach.

Nevertheless for all these installations the major hazards and their means of control should be identified. However this may necessitate a different approach to that taken for a future installation where FSA principles will be adopted from early concept development.

7.1 Strategy

It will be essential for all parties (regulatory authority, Operator, contractor(s), etc.) to have a common approach and understanding of FSA. Although the final regulations may be supplemented by guidelines, UKOOA has already produced guidance (Ref. 3) to assist a common intepretation between member companies.

Each operations management should establish its organisation with FSA in mind. To this end the early appointment of a safety professional practiced in hazard identification and assessment is likely to be essential.

The FSA should have the full support of the operating line management and not just be regarded as an additional increase in the Safety Department's scope of work. The balance of engineering and management controls presented in the Safety Case will commit the Operator to a certain style and depth of management, which will be audited periodically by the HSE.

7.2 Guidelines

The UKOOA guidance (Ref. 3) addresses the application of FSA to both new and existing installations. In particular the compatibility of the level of assessment on new platforms with that possible on older ones will be important.

There has been much discussion on the role of Quantified Risk Assessment (QRA) in relation to FSA. The preparation of a QRA, no matter how comprehensive, for a particular installation is NOT a FSA. FSA is a combination of many safety tools and techniques, addressing both engineering and management, one of which may be QRA.

The author believes it may be possible to present an acceptable Safety Case without the use of QRA at all. However there is one qualification to this statement. Lord Cullen has recommended (Ref. 2) the use of QRA to determine the risk to the integrity of the Temporary Safe Refuge (TSR), sometimes known as the 'safe haven'. The exact interpretation and approach to this aspect has yet to be determined by industry and the HSE, but may impose some mandatory requirement for QRA within the FSA.

The emphasis within the Safety Case will be on "structured qualitative reasoning" to demonstrate systematic hazard identification and the measures taken to prevent, control and mitigate the consequences of those hazards. As such the FSA is likely to be based upon engineering judgement, consequence analysis and QRA in that order.

The UKOOA guidance (Ref. 3) is likely to be extended in future to address the application of FSA to the full range of installations requiring Safety Cases. At present it specifically addresses production installations, although the general principles are more widely applicable. The International Association of Drilling Contractors (IADC) are also trying to develop a common understanding and approach between their members. Specific guidance will be developed in due course for mobile drilling installations, multi-purpose vessels, etc.

7.3 Scope

It is unnecessary for Operators of existing installations to produce a separate Concept Safety Case. However, an Operator may choose to discuss any significant concept engineering decisions within the core Safety Case.

As for a new installation, it will be necessary to demonstrate that all incidents which could present a Major Accident Hazard have been assessed and the appropriate action(s) taken. Therefore all such possible events must be identified and their potential causes understood.

A rigorous, formal method of hazard identification should be undertaken and fully recorded. Many approaches can be used to identify major hazards. One or more of the following techniques may be appropriate, depending upon the age of the installation and information available; the findings from HAZOPs, failure modes and effect analyses (FMEAs), audits, safety reviews, company and world-wide incident reports, etc. The use of check lists, such as that suggested in Section 4, is also an effective way of ensuring comprehensive identification.

The consideration of measures taken to protect against the identified hazards will be based largely upon 'structured qualitative reasoning' in line with UKOOA guidance. Engineering judgement backed up by consequence analysis should be sufficient in most cases. The emphasis of the argument should be upon hazard prevention, or, where this is not possible, upon demonstrating adequate control and mitigation of consequences. Where possible, findings from studies on related hazards on other installations should be incorporated to supplement the available information. Only on rare occasions should it prove necessary to resort to a comprehensive QRA of the installation. However it may be necessary to use a QRA approach to address the risks to the integrity of the Temporary Safe Refuge or safe haven.

The FSA may demonstrate that a particular hazard has not been adequately protected against. In this circumstance further studies will be necessary. These studies should investigate both engineering and management protective controls that could be implemented. Engineering modifications may not always be "reasonably practicable", and therefore it may be necessary to place more reliance upon management and procedural controls for existing installations. In demonstrating reasonable practicability it may be appropriate to use an ALARP approach with cost benefit analysis. It is implicit in this approach that engineering modifications, management and procedural controls, or their combination may not always be reasonably practicable. In the ultimate it may well be that it is not practical to do anything, and the risks exceed the Operator's criteria, in which case the Operator should consider a fundamental change in operation of the installation.

Where there is an emphasis on procedural controls, the Operator should demonstrate a sound understanding of the need for these measures throughout the organisation.

7.4   Timing

The preparation of Safety Cases retrospectively for the large number of existing installations represents a formidable task. Not only may Operators lack sufficient resources of suitable expertise, but there is evidence that this is widespread throughout the industry, ie. consultants, regulatory bodies and Certifying Authorities (CAs). Coupled with the experience of some of the larger Operators in meeting the onshore CIMAH regulations, a period of ca.5 years would not be unreasonable for the phased completion of studies of the depth and comprehensiveness required for all existing installations.

Each Operator should prepare a programme for undertaking FSAs on their installations, and agree this programme with the regulatory body. All Operators would be expected to show compliance by early submission of one or more Safety Cases.

7.5   Potential Problems

As stated above, the preparation of Safety Cases for new projects and retrospectively for existing installations represents a formidable task for many Operators. Not only may Operators lack sufficient resources of suitable expertise, but there is also evidence that this is widespread throughout the industry.

Some of the benefits that an Operator could accrue from a thorough review of its own activities could be lost if the major part of the FSA is not performed in-house. Indeed Lord Cullen drew attention to this issue in his report (Ref. 2). However, not all Operators will have adequate resources of the appropriate expertise to undertake this additional commitment. It is therefore essential that any consultants or contractors employed have a common understanding and interpretation of FSA.

The information available for some of the oldest installations will not be as comprehensive as that for a future installation, and the quality of that information is likely to be variable. Certain of the current techniques have only been developed and accepted offshore over the last few years, and are not as yet universally employed. Furthermore engineering standards may have advanced in the intervening period.

As stated above, the integrated nature of the FSA demands that a single body, competent in assessing both the engineering and management control aspects, should be responsible for its regulation. It will be exceedingly difficult, if not impossible, to separate the engineering and management controls. Furthermore the balance of these measures could vary significantly from one installation to another, whilst satisfactorily protecting against the hazards. It is therefore difficult to envisage a role for multiple authorities (DEn, CAs, Dept of Transport) as has been suggested in certain quarters.

7.6 Practical Experience/Lessons

Each Operator should establish an effective system of documentation control. The Safety Case will be a summary of relevant supporting documentation that may be subsequently called to justify the data and findings presented. To this end it is essential that management can demonstrate a comprehensive audit trail as to why engineering and operational decisions were taken, and that all actions raised have been resolved. All appropriate documentation should be retained by the Operator, who should be prepared to keep it for the life of the installation.

It will also be necessary to maintain this information up to date, for example P&IDs should be kept up to "as built" status to assist future hazard identification and avoid creating new hazards when modifying equipment or procedures. All design modifications should be subjected to a hazard identification technique, preferably HAZOP.

Experience to date has demonstrated that the hazard identification must be rigorous. It is insufficient to carry out a HAZOP of process P&IDs and a broad hazard analysis, such as commonly performed for Concept Safety Evaluations. A much more detailed approach is required that identifies, for example, all hydrocarbon inventories and all possible initiating events leading to loss of containment. Therefore diesel and ATK fuel should be considered besides the large hydrocarbon inventories associated with the reservoir, pipelines and process module(s). A similar level of detail should be applied to hazards other than hydrocarbons to identify all possible initiating events.

Problems have also been observed with uncontrolled changes to procedures. It is likely that existing installations will place more reliance on procedural and management measures to control certain hazards. In this instance it is vitally important that all concerned (managers, supervisors, operators and contractors) are aware of and properly understand these measures.

Whilst it is expected that many operations and maintenance personnel will input to the FSA, the final drafting of the Safety Case may best be undertaken by someone practised in hazard assessment. In this respect a safety professional able to relate the protective measures back to the hazards and their initiating events is appropriate. To date, such personnel have not always been employed by all Operators.

8. **SUMMARY**

FSA will demand a substantial effort and different approach to the safety management of the operation of existing installations. This in turn will require early management commitment and a clear strategy to meet the FSA objectives.

Nevertheless this new approach of a systematic and disciplined safety analysis offers substantial benefits, and is certainly seen as more beneficial to safety than the proliferation of prescriptive regulations.

9. **REFERENCES**

1. Offshore Installations - Formal Safety Assessments, Discussion Document by the Department of Energy, October 1989.

2. The Public Inquiry into the Piper Alpha Disaster, The Hon Lord Cullen, November 1990.

3. UKOOA Procedure on Formal Safety Assessment, UKOOA E+D Committee, Issue No. 1, November 1990.

TABLE 1

OFFSHORE SAFETY CASE

SUGGESTED CONTENTS

1. **Introduction**

   1.1 Regulations

   1.2 Installation

   1.3 Programme for FSA (may not be necessary for all existing installations)

2. **Corporate Safety Management**

   2.1 Company Safety Policy and Objectives

   2.2 Organisation - Head Office/Onshore

   2.3 Key Safety Responsibilities - Head Office/Onshore

   2.4 (Central) Support Functions - safety related

   2.5 Company-wide Safety Practices/Procedures/Standards

   2.6 Recruitment/Training

   2.7 Audits of Safety Standards and Practices

   2.8 Emergency Response

   Appendix   Safety Documentation Index (Company-wide)

3. **Description of Installation**

   3.1 General Description

   3.2 Location

   3.3 Environmental Conditions

   3.4 Layout/Configuration/Structure

   3.5 Manning and Personnel Location

4.　　**Installation Safety Management**

　　4.1　Organisation - responsibilities/onshore interface
　　4.2　Manning
　　4.3　Personnel - resourcing/training standards/competency
　　4.4　Control Procedures - planning/operations/maintenance/
　　　　　logistics/other
　　4.5　Safety Management Practices - implementation/inspection/
　　　　　training/audit
　　4.6　Emergency Response

5.　　**Information Relating to Potential Major Accidents**

　　5.1　Specific Safety Studies
　　5.2　Hazard Identification
　　5.3　Hazard Assessment - engineering judgement/consequence
　　　　　analysis/QRA
　　5.4　Safety Measures
　　　　　5.4.1　Major Accident Prevention
　　　　　5.4.2　Control Mechanisms
　　　　　5.4.3　Mitigation of Consequences
　　　　　5.4.4　Remedial Actions - programme
　　5.5　Justification for Continued Operation

6.　　**Company Verification Plan for FSA**

　　Appendix　　FSA Techniques and Data
　　　　　　　　　Safety Documentation Index (Installation Specific)

Can BS 5750 Satisfy Safety Management Systems?

A.Knights, Total Oil Marine, UK

Lord Cullen in his report on The Public Inquiry into the Piper Alpha Disaster[1] stated "I consider that operators should be required to set out formally the safety management system which they have instituted for their companies and to demonstrate that it is adequate for the purpose of ensuring that the design and operation of their installations and equipment are safe". He goes on further to say that ". . . I consider that in the formulation of their SMS operators should draw on principles of quality assurance similar to those contained in BS 5750[2] and ISO 9000".

The Report also notes in the context of Quality Management Systems, "I have come to the conclusion that it would be going too far for me to recommend the imposition of a system which would apply to all operators and across the entirety of their operations. I take the view that the operators should have the freedom to choose the type of system which is appropriate for them, in the light of the regime's requirements and their own operations." It is therefore quite clear that Lord Cullen recognised that at least a management system, as it relates to safety, using the principles of BS 5750, was necessary to control activities on an offshore installation. He did not feel that he was able to recommend a mandatory application of BS 5750 to all actions, rather allowing each operator the freedom to choose a system which was compatible to his activities.

Is it a reasonable assumption to make that the principles of BS 5750 applied to safety alone can in fact produce a Management System that will ". . . ensure that the design and operation of installations and equipment are safe", or is it necessary to consider the wider fields of management activities to satisfy the needs of safety?

This paper sets out to explore the application of BS 5750 principles to the production of a Safety Management System and tries to identify some of the potential pitfalls that will be found along the way. It will be shown that the System in itself is not a foolproof solution to satisfy the needs of ensuring safety, nor to produce the changes to achieve it; it will show that the behavioural aspects of the way the System is introduced and used are as, if not more, important than the System itself.

Lord Cullen proposed that the contents of a Safety Management System should demonstrate how the design and operation of our facilities should be controlled to improve safety and should include, as an example, the following:

"-organisational structure;
- management personnel standards;
- training, for operations and emergencies;
- safety assessment;
- design procedures;
- procedures, for operations, maintenance, modifications and emergencies;
- management of safety by contractors in respect of their work;
- the involvement of the workforce operators' and contractors' in safety;
- accident and incident reporting, investigation and follow-up;
- monitoring and auditing of the operation of the system;
- systematic re-appraisal of the system in the light of the experience of the operator and industry."

Can we infact consider safety as a function in itself? The Cullen Report already identifies that within a Safety Management System we should consider the elements of training, design, operations and maintenance. Similarly it is quite obvious that the activities of exploration, drilling, logistics and human resources must be involved. Without too much difficulty we have, therefore, identified that in the process of satisfying safety management we must look at virtually all of our corporate structure. Therefore, even if we only wish to identify safety as the function which we desire to improve we have no option but to provide a management system that covers the complete spectrum of our management activities. It has to be concluded that in applying the principles of BS 5750 to safety we have to consider the wider environment, what is generally called our quality management system.

It may seem that this is a different conclusion to the one reached by Lord Cullen, but in reality it is the same. A corporate-wide management system is required. That system will be different for each company, for each installation. There no set rules, except that safety management systems should have some degree of common approach across the industry. That approach being BS 5750.

BS 5750 already acknowledges this fact in Part 0: Section 0.2., Clause 1; "The extent to which these elements are adopted and applied by a company depends upon factors such as . . . nature of product, production processes . . ."

Having identified that what in fact we do require is a wide management system, can BS 5750 be applied to this process? Within the Guide to Quality Management and Quality System Elements (BS 5750: Part 0: Section 0.2., ISO 9004) it clearly states that this is a guide to "a basic set of elements by which quality management systems can be developed and impemented".

So not only do we have to look at the wider aspect of management systems, a part of BS 5750 is specifically designed to guide us to its application to the control of our management system.

BS 5750 is built around the application of twenty basic principles, and demonstrates to us in Part 0: Section 0.2. how to apply them. The elements we are suggested to consider are:

- Management responsibilty;
- Quality system;
- Contract review;
- Design control;
- Document control;
- Purchasing;
- Purchaser supplied product;
- Product identification and traceability;
- Process control;
- Inspection and testing;
- Inspection, measuring and test equipment;
- Inspection and test status;
- Control of nonconforming product;
- Corrective action;
- Handling, storage, packaging and delivery;
- Quality records;
- Internal quality audits;

- Training;
- Servicing;
- Statistical techniques.

It is not the intention to detail the requirements of each one of the twenty items above but simply to identify the relevance of them to the satisfaction of the previously stated content of a Safety Management System as stated by Lord Cullen.

Organisational Structure

As part of the overall Quality System it is clearly stated in Clause 5.2.3. (Section 0.2.):

> "The organisational structure pertaining to the . . . management system should be clearly established within the overall management of a company. The lines of authority and communication should be defined."

Management Personnel Standards

Although not specifically referred to in BS 5750 the needs to identify the standards to be set for management personnel are inherent as it is their responsibility to be the "owners" of the System, "The responsibility for and commitment to a quality policy belongs to the highest level of management", to provide motivation, to clearly delegate responsibilities for the management of the system as well as for all functional requirements.

Training, for Operations and Emergencies

Training is more than adequately referred to in BS 5750: Part 0: Section 0.2. clearly identifying the training objectives of all personnel, executive, technical and operational, emphasising that particular attention should be given to problem identification, analysis and corrective actions. Production personnel should be trained in the skills required to perform their tasks, including the equipment they have to operate and safety in their place of work. Formal qualifications as considered to be required in some instances, but consideration should also be given to both experience and demonstrated skills.

Safety Assessment

Within the area of statistical techniques the need to consider safety evaluations and risk analysis is idenfified, and indeed throughout the document safety is a common thread.

Design Procedures

BS 5750: Part 0: Section 0.2 more than adequately covers the area of design control identifying that: (as some examples)

> responsibilities should be defined, (Clause 8.2.1.);
>
> time phased design checks should be arranged to consider such aspects as unintended use and misuses, safety and environmental capability,compliance with regulatory requirements, failure modes and effects analysis, etc. (Clause 8.2.3.);
>
> the designer should give due consideration to the requirements of safety and environmental matters, (Clause 8.2.4.);
>
> account should be taken of reliability, and maintainability throughout the life expectancy of the design.

All the requirements of design control require to be documented such that adequate verification of the process can be carried out.

Procedures for Operations, Maintenance, Modifications and Emergencies

Substantial emphasis is made of the need to document the required management system in the form of policies and procedures to cover all areas and functions of the organisation. These should be set out with the target to co-ordinate different activities and should incorporate the objectives set by the organisation.

All these procedures should be written in a clear unambiguous form and should detail the requirements for verification and audit.

Specific procedures should be implemented to authorize the works necessary to implement modifications.

It is all very well identifying some simple generic terms that may form the content of the Safety Management System. From this point it is now necessary to provide some more substantial guidance on the subject. Tony Barrell, the new head of the Offshore Safety Division at the UK Department of Energy clearly identified this problem in a recent interview[3] noting that, ". . . there doesn't exist yet in a convenient form detailed guidance on how to set up a really good system". He follows this by noting that BS 5750 ". . . is in itself not enough as this doesn't say much about objectives".

The first of these two comments is being addressed by UKOOA who have established a Working Group to specifically identify the need for guidance and the form it should take. The second point is of particular interest and one which we should explore in more detail.

Although it would appear, either directly or by implication, that the principles of quality assurance as proposed by BS 5750 could easily satisfy the needs of a Safety Management System there a number of considerations which must be made to adequately understand the environment in which it has to be applied.

If we consider the elements of the proposed Safety Management System, it is very quickly noticed that a substantial number of the items; organisational structure, safety assessments, procedures for design and operations, accident and incident reporting, monitoring and auditing and re-appraisal refer to the application of rigid practices which in themselves can be deemed to be prescriptive. Some other items relate to the standards required of management and for training. Only one item actually requires the workforce to be involved in the process of ensuring safety. It can therefore be perceived that the safety management system is first of all prescriptive and is predominantly procedural in nature, all of which certainly outweighs the behavioural input of people.

It is being argued by the oil and gas industry and implied in the Cullen Report that prescriptive regulation will not assist in improving the way in which we operate and that we should be working in a goal-setting environment. Nevertheless the requirements of Safety Management Systems may, if applied too literally, not take us towards these objectives, leaving us steeped in procedures and reporting that entrenches us in the old-fashioned lowest common denominator mode.

Applying the principles of BS 5750 does not take us much further along the road to improvement. As has been demonstrated it can more than adequately satisfy safety management system needs and does indicate that objectives and policies are required around which procedures must be set. It highlights that motivation is an essential requirement. But still the predominant theme through the Standard is for procedures, reporting, inspection, checking, reviews, etc. In itself this is not wrong but it does lead to a rigid approach to the application of the principles and it can therefore be quickly forgotten that the major part in the application of any procedure is the people which are required to apply it. People do have the largest influence on the application of any system, but little real reference is given to the behavioural aspects required to really make the process of improvement and change work.

There are too many case histories of failed management systems. In a survey by Develin & Partners[4] published in March 1990 a review was carried out as to the success of the application of quality improvement programmes, and it identified quite clearly that;

almost all companies failed to achieve a change in management behaviour;
inadequate resources and time were given to the process;
very few tangible results were ever achieved as a result of all the effort.

Similarly the Harvard Business Review[5] in an analysis of a number of major United States companies identified similar problems:

a low success rate of change was actually achieved;
very low tangible results were produced despite considerable effort and cost.

The underlying problem in the effort to create change and improvement is that all to often no clear definition of the original business problem nor the objectives were actually idenfified. Great emphasis was put upon producing mission statements, corporate culture change, training programmes,etc., but with no clear targets.

Both reports came to the same conclusions as to what was required to create the required change and improvement. Primarily this involves identifying the basic business problem; improving inter-group communication, understanding and commitment; and, establishing the goals and the way to achieve them.

All of these basic requirements can of course be obtained without a structured formalised management system. It is at this point, it is argued, that having addressed the behavioural attitudes and obtained personnel commitment, that the management system is of use.

The (safety or other functional) management system can then be applied to support the software, the people, and act as a revitalising agent to maintain the momentum and enthusiasm created by the people who have to operate it. It will demonstrate that the system exists, it can be reviewed by regulatory authorities, and will form the basis for monitoring and auditing. Similarly this process enables the feed-back loop to be created to continually adjust and improve the way in which the system is working.

One area in which these two studies appeared to differ was the area of management commitment. Although it was agreed that top-down commitment was by far the best method for introducing change the Harvard Business Review suggested that as a first stage this may not be necessary. They noted that senior executives were by nature "cautious" of the effect that wholesale change could make to their organisations. Therefore it is suggested that individual departments or locations should lead in the process of change to demonstrate that success can be achieved. It is then argued that at this point the senior executive must take the process on board and expand its use throughout the organisation as a positive, successful theme of management.

The Harvard Business Review therefore suggests a six-point proposal to introduce change:

> Encourage commitment through the identification and solution of specific business problems,
>
> Develop a consensus of opinion as to the goals required and the ways in which this should be organised,
>
> Establish the manner in which the goals are to be achieved and the inter-group co-operation required,
>
> Generate a critical mass of personnel to create the momentum of change,
>
> Introduce the systemised methods to reinforce the change process and to maintain the momentum,
>
> Monitor and adjust the process continually in response to the identified problems,

It is therefore being proposed that the straight-forward implementation of a management system will not create the environment which will allow the system to work and subsequently create the requisite changes in our industry which we require. In fact this application of the system may cause more damage than benefit due to heightened expectations created within outside agencies (Department of Energy/ Health and Safety Executive) who may find on investigation that no real changes have in fact taken place. It is therefore essential to recognise that the major component required for successful improvement and change are the people who have to apply it. The system is necessary but should not overshadow, indeed it should only be the supporting star to the people, for it is only with their involvement, enthusiasm and commitment that we will achieve our goal to make our industry an improved one.

References

1. Cullen, Hon. Lord, *The Public Inquiry into the Piper Alpha Disaster,* 1990 (HMSO, London).

2. *BS 5750: 1987 Quality Systems* (British Standards Institution, London).

3. Knott, Terry, 'Manpower and money issues confront safety regulator', *Offshore Engineer,* January 1991, (Thomas Telford Ltd., London).

4. Develin & Partners, 'What's Wrong with British Quality', *British Quality Association: Newsletter,* March 1990.

5. Beer, M., Eisenstat, R.A., and Spector, B., 'Why Change Programs Don't Produce Change', *Harvard Business Review,* November-December 1990.

SESSION B:RESEARCH, RISK REDUCTION AND DESIGN SAFETY

Hyperbaric Ignition and Combustion Behaviour for some Selected Diving Chamber Specific Materials

H Boie, K Schmidt, A Tiemann

(GKSS-Research Centre Geesthacht GmbH, Germany)

ABSTRACT

Fire safety is one of the most important safety parameters which has to be considered for safe diving practice. The application of newly designed materials in the hyperbaric workplace environment calls for an experimental prove of the combustion behaviour of such materials. An experimental programme should demonstrate the limits between combustion and non-combustion for 8 textiles and 3 rubber/plastics, which will normally be used within an underwater welding habitat, dependent on oxygen content and diving depths. The test facility was operated with air and trimix gas mixtures.

The test results have been evaluated by comparison with the well known paper-strip experiments by Dorr (1971). The present test results show no more critical conditions relative to required minimum oxygen contents for hyperbaric combustion, therefore the paper-strip results remain a conservative and safe base for hyperbaric fire safety evaluation furthermore. A contamination of a textile with a combustible hydraulic oil decreases the non-combustion limit down to lower oxygen concentrations as expected.

1. INTRODUCTION

Typical diving conditions may include volumetric narrowness and normally limited possibilities to escape from the diving chamber or habitat. Therefore fire safety has to be considered with high priority in the discussion of diving safety conceptions. Only 10 years ago an official evaluation presented by Rosengren (1980) classified fire to be the main danger for the divers in chambers with 2.7 fatalities per event.

To ignite and to maintain a fire implies the existence of an ignition source, combustible materials and a sufficient O_2-content within the surrounding atmosphere. Fire safety precautions are directed to withdrawal of one of these preconditions, which can be difficult or impossible for hyperbaric diving activities such as dry welding in underwater habitats.

Normal practice depends on compromises faced with the following parameters of the hyperbaric welding environment:

- welding arc and welding spatter represent unavoidable ignition sources,
- material characteristics relative to combustibility have to be considered dependent on local ignition energy and oxygen percentage within the ambient atmosphere,
- oxygen contents within the diving chamber atmosphere amount to partial pressures greater than 0.21 bar for air diving and in the range between 0.25 and 0.5 bar for mixed gas diving normally.

Since in hyperbaric welding different combinations of materials and oxygen contents can be found, it is necessary to establish acceptable limits for a given fire risk.

With special view to the extended application of newly designed materials in the hyperbaric environment (e.g. materials for pressure hoses, electrical isolation, diver clothing) the Working Group "Underwater Welding" of the Deutscher Verband für Schweißtechnik (DVS) suggested to perform hyperbaric fire tests. The test programme should especially determine the limits between combustion and non-combustion as a function of oxygen content and diving depths. Moreover, Commission VIII, Working Group D of the International Institute of Welding (IIW) has considered hyperbaric fire aspects to be a matter of priority in the discussion of underwater welding safety in 1988. A reproduction of the actual state of the art in hyperbaric fire safety compiled for the IIW recommends further experimental work to avoid an undetected hyperbaric hazard potential.

2. AVAILABLE TEST RESULTS AND STANDARDS

A first extensive experimental programme was performed by Dorr (1971) after chamber fires with fatal consequences. Special attention was given to flammability and combustion of paperstrip specimen in helium-oxygen and nitrogen-oxygen atmospheres with different oxygen concentrations dependent on the total pressure within the gas volume. Until today these results represent a fundamental basis for the decision if an inflammable solid material meets fire safe conditions for specific oxygen contents and diving depths. Further basic experimental results were published by Rodwell et al (1985) regarding possible effects of hot welding spatter splashing on towel material in oxygen-nitrogen atmospheres with oxygen partial pressures between 0.15 and 1.2 bar for total pressures up to 6 bar. Further experiments concerning hyperbaric ignition and combustion of solid materials in diving specific atmosphere are not known.

Approved standard test procedures with reference to flame-resistance of textiles or to flammability of plastics, e.g. the NFPA Standard Methods, do not consider hyperbaric conditions. Likewise the well known Limiting Oxygen Index (LOI)-indicating the minimum oxygen concentration necessary to maintain the flame at combustion - is not applicable to hyperbaric conditions. All in all, the specification of standard conditions for ignition and combustion of solid materials in hyperbaric atmospheres is still pending.

3. EXPERIMENTAL PROGRAMME

The main purpose of the programme is to test materials which could be used within an underwater welding habitat relative to ignition and combustion behaviour as a function of oxygen concentration and diving depth. 11 different materials were considered for the present experimental programme, 8 textiles and 3 rubber/plastics. In the current study because of trademark protection, the trade names have not been mentioned. Further product information can be made available on special request.

3.1 TEST PROCEDURE

The test programme has been performed within a chromium-nickel pressure vessel (volume 40 ltrs., max. operational pressure 30 bar). Two acrylic glass windows allow observance of the combustion process until smoke and dust reduce the visibility within the vessel. For safety reasons there is a temperature control for the windows (max. temperature 60 °C). The tests were performed with air and trimix gas mixtures (5 % N_2, O_2 flexible from 4 to 21 % acc. to actual test parameter, remainder He).

The test specimens were fixed on a frame containing 8 pins, which can be moved out of the test vessel. 5 thermo-couples are installed to detect the temperature conditions at each specimen (1 thermo-couple placed near by the filament, 4 thermo-couples above).

Ignition heat was supplied by a plane coiled filament. The special nickel chrome alloy wire enables ignition temperatures greater than 1000 °C at max. power output of 250 W at 10 A. Between test specimen and filament was a distance of about 2 mm. The geometry of the test specimen is 65 x 35 mm. During all tests the specimens were fixed in an angular position of 45 degrees. This position should be a realistic compromise between horizontal position with max. heat accumulation before ignition and vertical position with max. flame spread after ignition. The tests by Dorr (1971) were performed with vertical positioned paper strips.

The time point of ignition and start of combustion was established by visual observation. Automatic thermo-optical procedures (e.g. flame guards) could not prove sufficient reliability because of varying flame appearance for different materials. After observation of flame formation through the acrylic glass windows the heat supply was immediately interrupted. After that the combustion could continue or the flame would extinguish, hence indicating incomplete combustion. To meet the attribute "non-combustible" a given material should withstand a 40 s continuous heat supply without ignition.

The experimental data has been recorded using a digital data acquisition system, which includes an amplifier, an analog-digital transformer and a PC with a software for data accumulation, handling and recording.

3.2 Test Results

According to good diving practice textiles or fabric of flame retarding quality would normally be used for hyperbaric working procedures. Therefore the selection of products for the fire tests has considered relevant materials with one exception, that is pure cotton which may be used as cleaning rag, in order to demonstrate the different combustion behaviour.

All experimental results have been set in diagrams presenting oxygen percentage in the gas atmosphere versus chamber pressure. The 0.5 bar oxygen partial pressure isobar is included for orientation to indicate an upper limit of oxygen compatibility for long-term breathing. At saturation diving - e.g. with trimix - breathable bottom gas mixtures as chamber atmospheres contain a max. oxygen portion corresponding to this partial pressure.

- Textile 1 (Fig. 1)

This product is a high temperature retarding fabric with increased pyrolysis temperature and a high percentage of noncombustible pyrolysis products. It will be used as protective clothes for divers during hyperbaric dry welding.

Some tests in pressurized air resulted in complete combustion (ignition temperatures 540 + 940 °C). This combustion however doesn't effect a flame spread around the area of the heated zone, it remains strongly restricted on the direct filament influenced area.

For trimix atmospheres the test results show no ignition for oxygen concentrations below 13.3 %. Max. heating temperatures for the pressure range 15 + 30 bar differ between 520 and 320 °C), the tests with trimix and pressures lower than 15 bar show max. heating temperatures between 955 and more than 1000 °C.

- Textile 1 contaminated with hydraulic oil (Fig. 2)

Hyperbaric work may include arc welding and the use of hydraulically operated tools simultaneously. A failure of the hydraulic system may cause a leakage of the hydraulic fluid into the hyperbaric environment. Although there are flame retarding or non-combustible fluids, it is not known which consequences should be expected in the case of combustible fluids. To answer this question the combination of Textile 1 with a hydraulic oil (flame temperature 210 °C) was tested relative to its combustion behaviour. The fabric test specimen soaked with oil was left to drip off in air for 24 h before the test started.

Ignition temperatures were determined to 880 °C in air at atmospheric pressure and to 320 °C at 10 bar. The tests in trimix atmosphere resulted in a broader range of incomplete combustion with ignition temperatures between 540 and 785 °C compared with the tests of the pure Textile 1. As expected, a clear decrease of the non-combustion area down to lower oxygen concentrations has been caused by contamination with this type of oil. Looking at the combustion, residues show a complete combustion of the contaminant but not of the fabric itself for all tests in trimix and the one in air at atmospheric pressure. Combustion in air at 10 bar resulted in simultaneous combustion of fabric and contaminant.

This test series demonstrated the remarkable decrease in safety as a consequence of an unexpected contamination of normally flame retarding materials.

- Textile 2 (Fig. 3)

Compared with Textile 1, Textile 2 has similar properties relative to temperature retardation. The product will be fashioned for special clothes, e.g. for divers too.

The results for pressurized air showed incomplete combustion at up to 10 bar (ignition temperatures between 420 and 1050 °C). All tests in trimix did not result in ignition (max. heating temperatures between 340 and over 1000 °C).

- Textile 3 (Fig. 4)

This material is used for the production of heat-resistant textiles for the industry, an application in the hyperbaric sector is unknown.

Tests have been performed in air atmosphere, the results show incomplete combustion up to 3 bar (ignition temperatures between 530 and 1030 °C) and complete combustion at pressures greater than 3 bar (ignition temperatures between 930 and over 1000 °C).

- Textile 4 (Fig. 5)

This Textile 4 is used as protective clothes for welders, an application for hyperbaric welding is unknown.

Tests in air atmosphere up to 10 bar resulted in complete combustion (ignition temperatures between 435 and 860 °C).

- Textile 5 (Fig. 6)

Fields of application are the same as for Textile 4. Tests were performed in pressurized air, the results show incomplete combustion up to 10 bar (ignition temperatures between 895 and 1040 °C).

- Textile 6 (Fig. 7,8)

This is a special heat retarding synthetic fabric which is applied in protection clothes for fire-fighting, aviation and aerospace. Often the pure Textile 6 is processed with other fibres, the tested mixture had a 60 % addition of a further fibre.

The pure Textile 6 (Fig. 7) does not burn in pressurized air (1, 6 and 10 bar, max. temperatures between 780 and 1090 °C) and in trimix (11 bar, more than 1000 °C).

For the mixed fibres fabric (Fig. 8) a reduced flame retardation was observed in pressurized air (incomplete combustion at 6 and 10 bar, ignition temperature more than 1000 °C). In trimix the mixture did not burn (19, 21 and 31 bar, max. temperatures between 520 and 660 °C).

- Textile 7 (Fig. 9)

This product is based on a natural fibre (cotton) which is specially treated for flame retardation. The fabric is offered with different weights per area, the test specimen had the max. value of 620 g/m^2. The product is offered for use as protective welding clothes.

The test results for pressurized air indicate incomplete combustion (1, 4, 6 and 10 bar, ignition temperatures between 680 and 1060 °C). For trimix atmosphere the transition from incomplete to non-combustion can clearly be recognized for system pressures greater than 10 bar. Incomplete combustion required ignition temperatures between 680 and more than 1000 °C, max. temperatures at points of non-combustion varied between 570 and over 1000 °C.

- Textile 8 (pure cotton) (Fig. 10)

The well-known fire sensitive behaviour of cotton could be demonstrated in hyperbaric trimix atmosphere. Roughly estimated flame spread rates increased with rising oxygen contents or partial pressures respectively. Ignition temperatures have been measured in the range of 820 °C to more than 1000 °C.

- High pressure hose (Fig. 11)

For reasons of breathing protection the hyperbaric welding chambers are equipped with built in breathing systems which consist of mask, regulator, supply and disposal hoses. When operating several sets with sufficient hose lengths a remarkable fire load would be piled up within the chamber. The composition of the hose material varies (india rubber, polymers, soot and different oils), therefore the following results should not be regarded as representative.

The tested material shows a combustion behaviour safer than expected, the area of incomplete combustion remains limited for oxygen concentrations greater than 12 % (ignition temperature between 280 and 550 °C).

- Low pressure hose 1 (Fig. 12)

A polyvenyl chloride hose is used as a flexible pipe to exhaust welding fumes from the work place to the gas absorber.

The material ignites for incomplete combustion at relative high oxygen concentrations too (ignition temperatures between 165 and 195 °C).

- Low pressure hose 2 (Fig. 13).

A polyethylene hose is very light and flexible (comparable to a vacuum cleaner hose) and is used for special breathing sets presently in development.

The test results showed discriminating areas of combustion (ignition temperatures between 100 and 340 °C).

3.3 EVALUATION AND OUTLOOK

The excellent work by Dorr (1971) includes a scale of fire resistance, which applies non-flammability and self-extinguishment of materials in atmospheres with oxygen concentrations between 21 % (air) and 100 % pure oxygen. The scale is divided in 10 classes, starting with class 0 (burns readly in air at atmospheric pressure) and finishing with class 9 (non-flammable in 100 % oxygen at a pressure of 1 bar). This scale however does not sufficient consider typical oxygen concentrations for trimix saturation diving, e.g. 0.5 bar partial pressure at 100 m diving depth corresponding to 4.5 % oxygen concentration. Consequently the definition of fire resistance classes for oxygen concentrations lower than 21 % is still pending.

For evaluation of the present test results the results of the paper-strip flammability tests performed by Dorr (1971) will be considered for comparison. Dorr's (1971) experiments in heliox mixtures cover a range of oxygen between 2 and 24 % for pressures between 1 and 43 bar.

Compared with the paper-strip experiments the existing material test results show no more critical results relative to required minimum oxygen contents for hyperbaric combustion. For this reason it seems conservative and safe to consider the paper-strip results as a base for hyperbaric fire safety decisions as before.

The fire safety experimental programme at GKSS is a on-going project. It is foreseen to condense the number of measuring points for some selected materials along the 0.5 bar oxygen partial pressure isobar for system pressures lower than 5 bar. Furthermore there is an interest in the installation of some gas analysis equipment, because the hyperbaric fire will lead to the formation of toxic gases. Another question relates to possible consequences of fire induced heat impacts on the acrylic glass windows of a pressurized diving chamber.

ACKNOWLEDGEMENTS

The authors appreciate the work of Mr. M Döring and Mr. B Reimers, both were involved in the performance of the test programme in the frame of working out their graduation theses. Many thanks to Mrs. E Lieder who has typed this paper.

REFERENCES

Dorr, V.A. (28 Feb. 1971) Compendium of Hyperbaric Fire Safety Research; Final Report on Combustion Safety in Diving Atmospheres, Ocean Systems, Corp./Union Carbide Corp., Tarrytown/New York

Rodwell, M.H. and Moulton, R.J. (July 1985), Fabric Flammability under Hyperbaric Conditions, The Welding Institute, Abington, Cambridge CB1 6AL.

Rosengren, P. (1980), 'Lessons to be learned', Technical and Human Aspects of Diving and Diving Safety, International Symposium of the Commission of the European Communities, Directorate for Health and Safety, The European Diving Technology Committee, Doc. n. 3000/80 E, pp. 49.

International Institute of Welding (1990), Draft on the State of the Art Report Hyperbaric Fire Safety, Working Group VIIID, London.

National Fire Protection Association (1977), Standard Methods of Fire Tests for Flame-Resistant Textiles and Films, NFPA 701-1977.

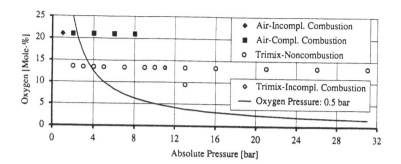

Fig. 1 Combustion behaviour Textile 1

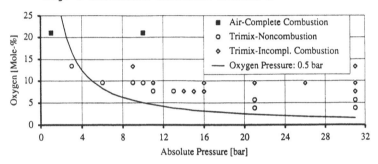

Fig. 2 Combustion behaviour Textile 1, contaminated

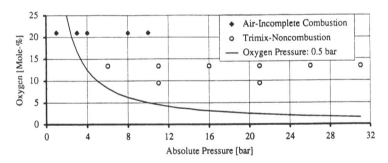

Fig. 3 Combustion behaviour Textile 2

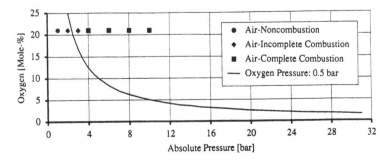

Fig. 4 Combustion behaviour Textile 3

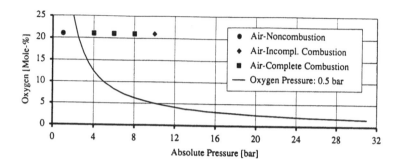

Fig. 5 Combustion behaviour Textile 4

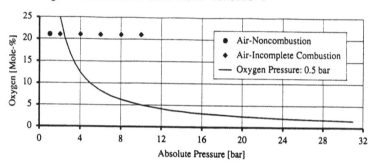

Fig. 6 Combustion behaviour Textile 5

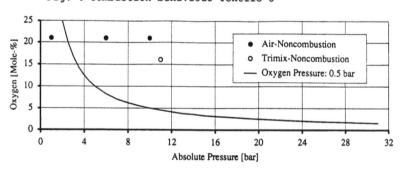

Fig. 7 Combustion behaviour Textile 6, pure fibres

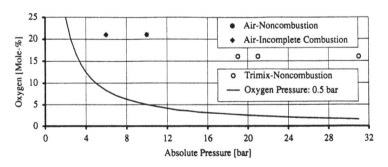

Fig. 8 Combustion behaviour Textile 6, mixed fibres

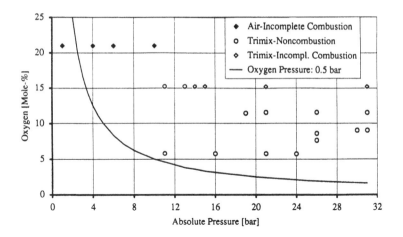

Fig. 9 Combustion behaviour Textile 7

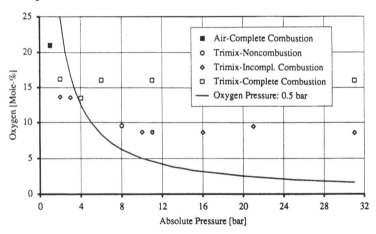

Fig. 10 Combustion behaviour pure cotton

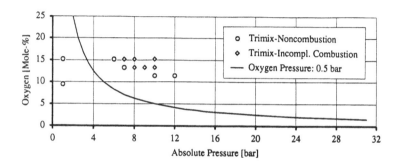

Fig. 11 Combustion behaviour high pressure hose

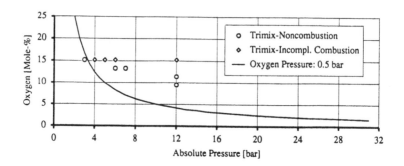

Fig. 12 Combustion behaviour low pressure hose 1

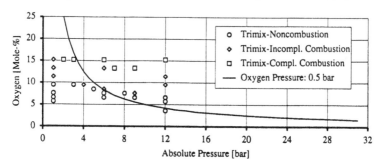

Fig. 13 Combustion behaviour low pressure hose 2

MODELLING OF MISSILE ENERGY FROM GAS EXPLOSIONS OFFSHORE

Dr V H Y Tam & Miss S A Simmonds
BP Research

ABSTRACT

In gas explosions on offshore platforms, very high gas flow velocities can be generated. Although this high flow only lasts for a short time, it can impart a significant amount of kinetic energy to loose or poorly secured objects. These flying objects, or missiles, could cause further damage to equipment and loss of containment. In addition to missile characteristics, eg. its shape, the potential for damage depends upon the efficiency with which energy from the gas explosion is transferred to missiles. In this paper, a method to calculate the maximum speed of missiles is described. This is illustrated with examples using objects commonly found on offshore platforms, such as scaffold clips, and using gas velocities calculated from numerical simulations of gas explosions on offshore platforms.

INTRODUCTION

The damaging consequences of gas explosions in major accidents, such as those in Flixborough and Piper Alpha are clearly large in terms of financial and human cost. In order that the consequences of gas explosions can be mitigated, a basic scientific understanding of the physical processes in gas explosions is necessary.

A gas explosion could cause direct damage to structures or large equipment via the high overpressure wave it generates. Much research effort in the past decade has been on the combustion-fluid flow interactions in gas explosions. As a result, predictive tools for estimating overpressures from gas explosions are now available (Hjertager (1985) and (Catlin (1990)). Some of these models are already being applied to aid platform designs (DEn (1989), Tam and Simmonds (1990) and Bakke et al (1990)). A gas explosion could cause damage indirectly, by producing high energy projectiles or missiles (Baker (1975)). This is an area where little research has been carried out.

This paper describes a simple method to estimate the kinetic energy of missiles based on results from numerical simulations of gas explosions in realistic (but hypothetical) offshore platform designs. The calculated kinetic energy acquired by a number of example 'loose objects' are presented. A simple formula has been derived for this calculation. Example loose objects include a hard hat, and two types of scaffold clips.

GENERATION OF MISSILES

Missiles are frequently generated in gas explosions and they can cause the escalation of an accident. The ability of a missile to cause significant damage depends upon its kinetic energy on impact. Here, we address the issue of missiles generated by the fast gas flows occurring in gas explosions. We do not include missiles generated by physical explosions, eg caused by vessel rupture, or wall failure, eg. large pressure difference across a wall could cause hatches to be released.

There are two main mechanisms by which energy can be imparted from the gas explosion to the missile. These are via overpressure and drag forces.

1. Overpressure
The overpressure waves generated by a gas explosion give rise to net forces on loose objects which are of comparable size to the length scale of the pressure waves which is typically several metres. The mechanism involved in this case is reflection and defraction. However, for small objects, the forces generated by these effects are small. For the example objects used in this study, this effect is assumed to be negligible.

2.   Drag

Gas explosions also produce high air velocity and this produces a drag force on an object lying in its path. The drag force is expressed as:

$$F = C_d A (0.5 \rho V^2) \qquad\qquad (1)$$

where
F is the drag force in N
A is the area normal to the flow in square metre,
ρ is the density of the air in kg m^{-3}
V is the relative velocity of air and the object in m s^{-1}, and
C_d is the drag coefficient.

The force (F) is due to shear stress on the object and it varies with the level of turbulence the object generates. Thus the value of the drag coefficient C_d varies with flow speed, density and viscosity of the fluid, and the shape of obstacles. For steady flow, this problem has been extensively studied, and the values of C_d are available for a large range of Reynolds number and obstacle shapes.

TRANSIENT EFFECTS

In addition to the high gas velocity, other properties of the gas also change rapidly. For example, over a small fraction of a second, the density is reduced by a factor of 7 over the passage of the combustion front and air velocity can increase to several hundred metres per second. The way gas velocity and temperature coupled together will determine the total kinetic energy that will be transferred to the missile. Unfortunately, there is a lack of experimental data in this area.

1.   Velocity and Density Variation with Time

In order to generate information on velocity and density vs time curves, we used the FLACS code (3) which is developed by the Christian Michelsen Institute and sponsored by a consortium of companies including BP. Figure 1 shows the calculated velocity and density vs time curves for one location on a hypothetical platform. They were obtained from a numerical simulation of a gas explosion on a platform the layout of which is shown schematically in Figure 2. The curves in Figure 1 apply to location A in Figure 2. The peak velocity and its duration vary depending upon the location inside the offshore platform as well as upon the platform layout. For example, the maximum velocities at point B and C in Figure 2 are about 300 m s^{-1}. However, the shape of curves in Figure 1 is typical of velocity and density vs time curves calculated in a number of numerical simulations near the vent area.

There are two main features of velocity and density vs time curves in locations where the gas velocities are high (ie greater than 100 m s^{-1}):

(a) The coupling between velocity and density is low. Typically, the gas velocity begins to rise soon after the density has begun to fall. However, gas density falls relatively quickly to its minimum value long before the gas velocity has reached its peak value.

(b) The rise and fall in velocity is roughly linear with a very short period during which the velocity is at its maximum value.

2.   Idealised Velocity and Density-Time Curves

Idealised density and velocity vs time curves are shown in Figure 3 and incorporate the main features of those observed in numerical experiments. In the idealised situations, the gas velocity rises linearly with time until it reaches a maximum value, and falls linearly afterwards. The magnitude of acceleration and deceleration is the same. Further, the gas density is assumed to have fallen to its lowest value before the gas velocity starts to rise.

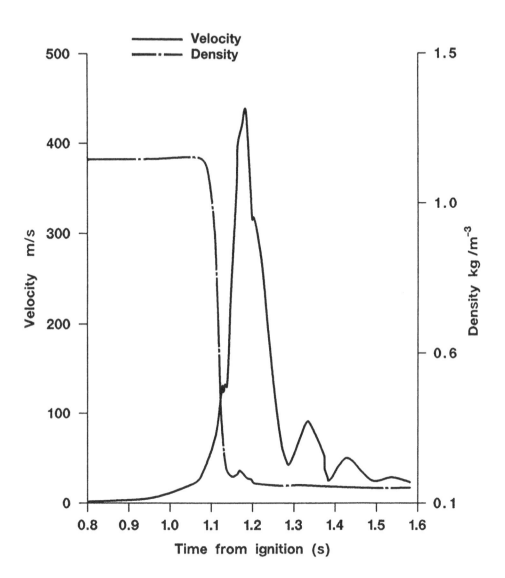

Figure 1. Velocity and density variation with time from a numerical simulation of a gas explosion.

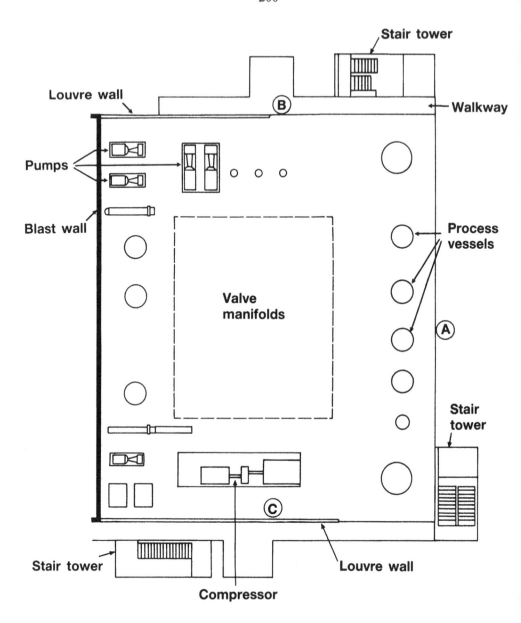

Figure 2. Schematic diagram showing the layout of a process level which is open on 3 sides. The points where the gas velocities are referenced in the text are also shown.

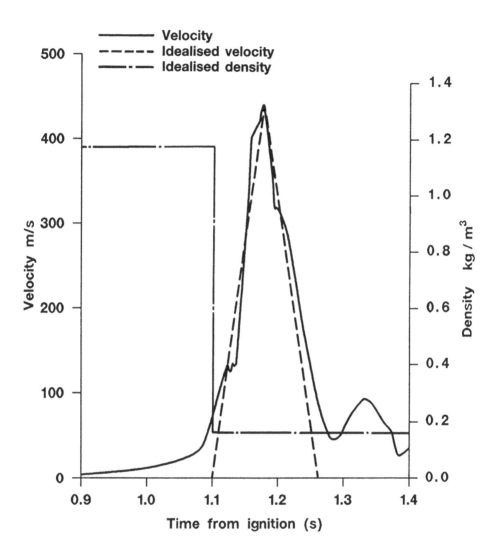

Figure 3. Idealised velocity and density variation with time based on numerical simulations of gas explosions.

VELOCITY OF MISSILES

The velocity of the missiles can be easily calculated using equation (1) and the idealised velocity-time curve (Figure 3). The gas velocity (V_g) is given by:

$$V_g = 2 t \qquad\qquad \text{for } t > 0 \text{ and } < 0.5$$

$$V_g = 2 (1 - t) \qquad\qquad \text{for } t > 0.5 \text{ and } < 1 \tag{2}$$

where

V_g = (Gas velocity)/V_{gmax}
t = time from when velocity starts to rise / t_{max}
t_{max} = the width of the velocity-time curve
V_{gmax} = peak gas velocity

Expressions for the missile (V_m) are obtained from equation (1) by substituting ($V_g - V_m$) for the relative velocity (V) and the above relationship for V_g, and integrating with respect to time assuming a constant gas density. The velocity of the missile (V_m) for the following three time periods are:

Phase (a): When gas velocity is accelerating, ie $t <$ or $= 0.5$.

$$V_m = 2t - \frac{\sqrt{2}}{\alpha} \tanh \left[\alpha \sqrt{2}\, t \right] \tag{3}$$

Phase (b): When gas velocity is decelerating, ie $t > 0.5$

$$V_m = 2 - 2t - \frac{\sqrt{2}}{\alpha} \tan \left[\frac{\alpha}{\sqrt{2}}(1 - 2t) + \text{arc } \tan(\tanh \frac{\alpha}{\sqrt{2}}) \right] \tag{4}$$

Phase (c): When the gas velocity has dropped below that of the missile

$$V_m = 2 - 2t + \frac{\sqrt{2}}{\alpha} \tanh \left[\frac{\alpha}{\sqrt{2}} (1 - 2t) + \text{arc } \tan (\tanh \frac{\alpha}{\sqrt{2}}) \right] \tag{5}$$

where V_m = velocity of missile / V_{gmax}
α^2 = 0.5 C_d ρ A V_{gmax} t_{max} / M
ρ = density of the gas
A = cross sectional area of the missile
M = mass of the missile

The missile is experiencing forces which are increasing with time during phase (a). As the drag force depends on the velocity difference between the gas and the missile, the missile will never attain the maximum gas velocity (V_{gmax}), ie V_m is always less than 1. This can be shown by deriving the maximum missile velocity (V_{mmax}), equation (6) below, using the equation (4) above.

Equation for maximum missile velocity

$$V_{mmax} = 1 - \frac{\sqrt{2}}{\alpha} \text{ arc } \tan \left[\tanh \frac{\alpha}{\sqrt{2}} \right] \tag{6}$$

EXAMPLES: APPLICATION OF THE MISSILE VELOCITY MODEL

As an example to illustrate the effect of gas explosions on missile velocity, we have made use of the velocity-time curve in Figure 1 to estimate the velocities of a hard hat and 2 types of scaffold clips. The idealised version of the density and velocity-time curve in Figure 1 showed that t_{max} = 160 ms, V_{gmax} = 440 m s^{-1}, and the gas density = 0.16 kg m^{-3}.

The results showing the velocities of the above objects are listed in Table 1. The trends are as expected; light large objects achieve high velocities whereas heavy small objects achieve relatively low velocities. In the case of the hard hat, its maximum velocity is over 100 m s^{-1}, containing 2 kJ of energy. Whereas the small heavy scaffold clip has a maximum velocity of 14 m s^{-1}, and gain 140 J of kinetic energy. The velocity-time curves for both the hard hat and the heavier scaffold clip are shown in Figure 4.

TABLE 1 Maximum Velocities of Different Loose Objects

Objects	C_d	Area (m^{-2})	Mass (kg)	α	Maximum Velocity (m s^{-1})	Distance to 90% Max Velocity (m)
Hard hat	1.4	.05	.37	1.04	105	2
Scaffold Clip 1	1.2	.009	.92	0.26	10	0.3
Scaffold Clip 2	1.2	.010	.69	0.30	14	0.5

DISCUSSION

i   Maximum Missile Velocity

The maximum velocity of the missile depends only on one single parameter α which can be thought of as a ratio of impulse to inertia per unit cross sectional area. The relationship between maximum missile velocity and α is shown graphically in Figure 5. α is large for large peak gas velocity, large duration of the velocity pulse, and low density objects. Conversely, it is small for small peak gas velocity, small duration of velocity pulse, and dense objects.

This can be illustrated using the hard hat in the above example. The value of α in the conditions described in Figure 1 is 1.04 and this gives a scaled maximum velocity (V_{mmax}) of 0.24. If it is exposed to a more violent gas explosion, say V_{gmax} = 750 m s^{-1}, T_{max} = 300 ms, the value of α would be increased to 1.87. The corresponding maximum velocity of the hard hat is 350 m s^{-1} (V_{mmax} = 0.46). Under similar conditions, the lighter scaffold clip is expected to have a maximum velocity of 69 m s^{-1}.

ii   Energy of missiles

The energy transferred to the missile depends upon the value of α. The higher the value of α, the greater is the energy transferred to the missile. For a given velocity-time curve, light objects will gain more energy than dense objects. This

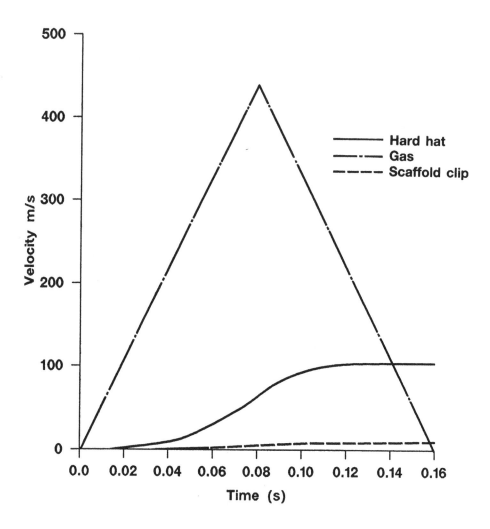

Figure 4. Velocities versus time for a (1) Hard hat, (2) Scaffold clip and (3) Gas.

can be illustrated using the above example of a hard hat and an additional hypothetical heavy hard hat. The heavy hard hat is 12 times heavier than the light hat. Otherwise, they are identical. The α values for the light hard hat and the heavy hard hat are 1.04, and 0.3 respectively, and their corresponding kinetic energies are 2 kJ and 0.9 kJ. Using the higher value of α of 1.87 for the hard hat and 0.57 for the heavy hard hat for the more violent conditions described in (i) above, the kinetic energies are 22 kJ and 11 kJ respectively. The corresponding kinetic energies of the lighter scaffold clip in the two explosion examples above are 0.1 kJ and 1.6 kJ respectively. Both maximum missile kinetic energy and velocity can be obtained from Figure 5.

iii Assumptions Made

In developing the methods above, we have made a number of assumptions. As some of them are based on a limited number of numerical experiments, they need to be verified, preferably with experimental data. The assumptions made are:

a   Drag Coefficient (C_d)
The drag coefficient in a steady flow is also applicable to that in transient flow conditions. Measurements will be required to verify this assumption.

b   Idealised Velocity-Time Curve
The idealised velocity-time curve is a true representation of velocity-time relationship in a gas explosion offshore. As our idealised curve is based on a limited number of numerical experiments, further data is needed.

c   Density-Velocity Coupling
The density-time curve is not coupled with the velocity-time curve.

d   Velocity-Time Variation As Seen by the Missile
The position of the missile has not changed significantly during the passage of the velocity wave. The velocity-time curve measured at one location is therefore representative of that seen by the moving missile.

The effect of this assumption (d) can be assessed. Using the example of the hard hat above, we can see from the velocity-time curve that the hard hat has gained 90% of its maximum velocity after 100 ms when it has moved about 3.2 m, see Figure 6. This is much less than the 9 m the missile would have travelled during the 160 ms. For missiles with lower value of α, the distance travelled would be even less, eg the heavy scaffold clip in our example would have travelled about 0.3 m before gaining 90% of its maximum velocity. Taking the more violent explosion as an example, the hard hat and the scaffold clip are estimated to have moved 17 m and 4 m respectively.

Obviously, the validity of this assumption depends upon the initial location of the missile (ie whether it is close to a vent), and its mass, shape and size. If the missile has not moved significantly during the passage of the velocity wave, the method used is likely to slightly underestimate its final velocity.

CONCLUSIONS

This paper has presented a method to allow a first estimation of the velocity and energy of missiles based on a single parameter in an offshore environment. We showed that light objects gain more energy than denser objects. In our examples, the hard hat attains a velocity of about 100 m s^{-1} (2 kJ) while a small scaffold clip of .7 kg achieves a velocity of about 14 m s^{-1} (0.14 kJ). In the example with a more violent explosion, the corresponding values are 350 m s^{-1} (22 kJ) and 70 m s^{-1} (1.6 kJ) for the hard hat and scaffold clip respectively. A number of assumptions have been made during the development of this method. These may affect its application. Experiments are needed to validate the theory, and in particular, the validity of assumptions used.

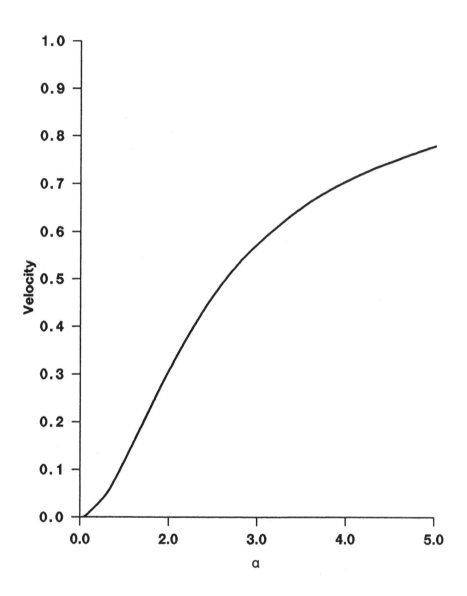

Figure 5. Maximum missile velocity (V_m) versus α.

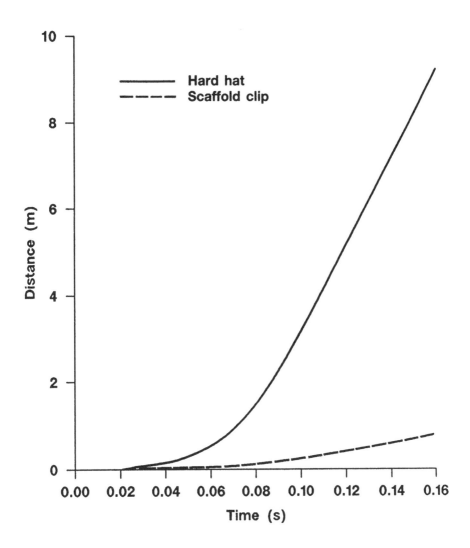

Figure 6. Distance travelled with time for a (1) Scaffold clip and (2) Hard hat.

208

ACKNOWLEDGEMENT

The authors are grateful to comments from members of the Safety Physics Group at
BP Research, Safety and Loss Control Department at BP Exploration and Corporte Safety
Services at BP and Dr Bakke from CMI.

REFERENCES

1. Hjertager, B.H. (1985), 'Computer Simulation of Turbulent Reactive Gas
    Dynamics', Modelling, Identification and Control, Vol 5, No 4, p221.

2. Catlin, C., 'CLICHE - A Generally Applicable and Practicable Offshore Explosion
    Model', I Chem E Symposium Series No 122, September 1990.

3. Department of Energy, UK, (1989), 'Review of the Applicability of Predictive
    Methods to Gas Explosions in Offshore Modules', Offshore Technology Report OTH
    89 312.

4. Tam, V.H.Y., and Simmonds, S.A. (1990), Effects of Equipment Layout and Venting
    Geometry on the Consequences of a Gas Explosion, Europec 90, Society of
    Petroleum Engineers, Hague, Netherlands, October 1990.

5. Bakke, J.R,, Bjerketvedt, D., and Bjorkhang, M. (1990), 'FLACS as a Tool for
    Safe Design Against Accidental Gas Explosions', in Piper Alpha - Lessons for
    Life - Cycle Safety Mangement, I chem E, London, September 26-27.

6. Baker, W.E., Kulesz, J.J., Ricker, R.E., Bessey, R.L. And Westine, P.S. (1975),
    Workbook for Predicting Pressure Wave and Fragment Effects of Exploding
    Propellant Tanks and Gas Storage Vessels, NASA Report No: NASA-CR-134906; REPT
    02-4130, November 1975.

LIST OF NOMENCLATURE USED

Symbols	Description
V_g	(Gas velocity)/V_{gmax}
t	time from when velocity starts to rise / t_{max}
t_{max}	the width of the velocity-time curve
V_{gmax}	peak gas velocity
V_m	velocity of missile / V_{gmax}
α^2	0.5 C_d ρ A V_{gmax} t_{max} / M
ρ	density of the gas
A	cross sectional area of the missile
M	mass of the missile

<u>LARGE-SCALE NATURAL GAS AND LPG JET FIRES AND THERMAL IMPACT ON STRUCTURES</u>

by

L.T. Cowley

Shell Research Ltd., Thornton Research Centre, P.O. Box 1, Chester CH1 3SH U.K.

and

M.J. Pritchard

British Gas plc., Midlands Research Station, Wharf Lane, Solihull, West Midlands B91 2JW, U.K.

1. <u>AVAILABLE JET FLAME DATA AND REQUIREMENTS</u>

The safe design and operation of natural gas and LPG facilities requires an ability to predict hazard consequences reliably. A particular hazard is a jet fire, sometimes called a torch fire, that might arise from the ignition of an accidental release of pressurised gas or liquid.

On offshore gas and oil production platforms and also on land based gas facilities, accidental releases might occur of high pressure natural gas, sometimes containing higher molecular weight components. Pressurised releases of two-phase condensate may also occur. Examples of potential release sources are the fracture of small to medium sized pipes and the resulting mass flow rates of high pressure gas are of the order of 0.3 to 20 kg s^{-1}. Important factors are flame extent, duration, external radiation, the degree of impingement onto vessels and structural members and the magnitude of the heat fluxes to flame impinged objects.

Accidental releases of pressurised hydrocarbons leading to jet fires might also occur at LPG facilities. Heating of a vessel by a jet fire produces an increase in internal pressure and concomitant weakening of the containing walls with the potential for escalation to a BLEVE. LPG releases may be purely gaseous or a two-phase flashing discharge of vapour and liquid droplets. The release rates of concern are again of the order 0.3 to 20 kg s^{-1} but at lower pressures (7 to 20 bar) than those from natural gas discharges. Knowledge is needed of flame extents, the effect of buoyancy and wind, the degree to which flames might engulf nearby vessels and the resulting heat fluxes to them.

To summarise, there is an industry need for data on the properties of LPG and high pressure natural gas jet fires in the flow range 0.3 to 20 kg s^{-1}. Flame impingement data and heat fluxes are of particular concern and are given most emphasis in this paper.

Soundly based information on the heat transfer from large jet fires to engulfed objects is almost completely lacking. Jet fires are commonly thought to have localised effects and estimates of the total heat flux have ranged from 50 up to 1000 kW m^{-2}. Steel structural members exposed to the latter value could fail within a few tens of seconds. The flux estimates are based on measurements in laboratory flames and sometimes confuse heat transfer in the free turbulent diffusion flames of interest here with those in premixed flames and in furnaces.

There are considerable problems in extrapolating small scale flame measurements to large turbulent diffusion flames because of the complex way in which the internal flame radiation and its distribution changes with flame size and other parameters. Flame radiation arises from molecular emissions and from soot. Molecular radiation (from hot H_2O and CO_2) dominates in small flames whereas in those of interest here, soot radiation may be significant or even dominant.

In flames having reasonably low soot volume fractions (e.g. natural gas flames) the path lengths for thermal soot radiation are several metres and hence these flames become more radiative with increasing size until optically thickness is approached. Whilst fundamentally based theoretical models are beginning to address these problems, the only method that is currently reliable for obtaining impingement heat fluxes is direct measurement in flames of adequate size. The only full size jet fire impingement data currently available is that on two-phase LPG releases of 1 to 10 kg s^{-1} in which limited impingement data on small targets was obtained and incident fluxes of 50 to 250 kW m^{-2} were measured[1,2].

In response to the needs for full scale experiments, Shell Research Ltd and British Gas plc have jointly conducted an extensive research programme to obtain directly, comprehensive information on the properties of natural gas and two-phase propane jet fires. The programme was co-funded by the Commission of the European Community under Contract EV4T-0016-UK.

2. THE EXPERIMENTAL PROGRAMME

A set of three 'standard' ignited two-phase propane releases and four natural gas releases were selected. Propane discharges were at a nominal pressure of 9 bar absolute with flow rates from 1.5 to 22 kg s^{-1} from apertures of 10 to 50 mm diameter. The natural gas releases had static discharge pressures from 1.2 to 58 bar absolute and flow rates of 3 to 10 kg s^{-1}, the lowest pressure discharge was sub-sonic and the others sonic.

The experiments were conducted using a purpose designed jet fire research facility built at the British Gas Test Site at Spadeadam, Cumbria, UK. The releases were horizontal with precise control of the flow rate and discharge conditions. Flame sizes, geometries, spot surface emissive powers, flame spectra and radiation to the surroundings were measured. In most tests the flames were directed against one of two alternative structures, a 13 tonne cylindrical LPG tank or a 900 mm diameter natural gas pipeline section. Heat flux distributions over the structures were obtained directly using an array of 40 fast response total heat flux sensors located on their surfaces. The radiative flux to the impingement targets and the target metal temperatures were also measured. The direct heat flux measurements were supplemented by measurements within the flame of gas temperatures and velocities. Experiments were conducted with source to target distances from 9 to 40 m. Figure 1 shows the release pipework, the 2 m diameter instrumented LPG tank target and instrumentation to measure the internal flame properties.

3. FINDINGS AND DISCUSSION

170 valid experiments were conducted and the available data are very considerable. Interpretation and analysis is continuing and this paper presents a summary of the firm findings to date of foremost interest to the industry.

3.1 Flame shape and external radiation

The two-phase propane jet fires were of low initial velocity and very buoyant with strongly curved and wind influenced flame trajectories. Radiation to near field objects and the potential for flame impingement are much influenced by these factors. As an example, whilst the largest propane flames were some 40 m in length as measured along the flame trajectory, the greatest distance at which impingement on the targets occurred was 21 m. The sub-sonic natural gas flame was also quite buoyant and wind influenced, whilst wind speed had little effect on the tilt and flame length of the ignited sonic natural gas releases.

The propane and natural gas flames had different radiative characteristics. The natural gas flames were less radiative with effective spot surface emissive powers varying considerably along the flame trajectory. The radiation was primarily from molecular emissions rather than soot and hence the flames were not optically thick at all wavelengths,

Figure 1. Propane jet flame impinging on an instrumented
13 tonne LPG tank 9m from the discharge point

with implications for prediction of near field and internal radiation and the influence of scale. In contrast, the propane flames were more radiative with continuum soot emissions dominant.

Current models[3,4] to predict the external radiation field around flares or jet fires are based on field scale jet fires using vertical or inclined flares with varying extent of cross wind. The current experiments together with previous ones on horizontally directed propane jet fires[1,2] provide a basis for improving predictive abilities in the regime where buoyancy is more dominant and the cross wind vector is differently oriented.

3.2. Flame engulfment and thermal impact

Jet flame impingement was not localised. Figure 1 shows the 13 tonne LPG tank target 9 m from the release point and the very substantial engulfment by a 12 kg s^{-1} propane flame originating from a 50 mm pipe orifice. This behaviour was similar for both the propane and sonic natural gas flames. The smallest releases (1.5 kg s^{-1} propane) still covered a significant area with flame wrapping around the targets.

The measured heat flux distributions reflect the non local impingement. Results for a sonic gas flame from a 75 mm dia. orifice, an upstream static pressure of 12 bar absolute and a mass flow rate of 8 kg s^{-1} are used as examples but the findings are generally applicable. The upper section of Figure 2 shows the total heat flux distribution over the pipeline target at a source distance of 9 m - the pipe is shown as a development (i.e. unrolled). The maximum heat flux of 250 kW m^{-2} at this particular source distance is on the rear of the pipeline and results from a combination of radiation from the substantial downstream flame and a convective flux of 160 kW m^{-2}. The immediate impingement (stagnation) point on the target front is in relatively cool partially burned gas giving a lower convective flux and a small radiative contribution. This flux distribution is one example and the distributions depend on source to target distance and source conditions.

Flux distributions from several tests are summarised in Figure 2 in terms of the target area exposed to a flux greater than a particular value. Flame impinged areas have incident total fluxes ranging from 50 to 300 kW m^{-2}. The peak flux is lower than most previous estimates and is effective over a small area of target and locality in the flame. The heat flux range is similar for all the sonic natural gas flames studied here.

The propane flames showed broadly similar behaviour with a cool core giving low heat fluxes near the stagnation point for targets placed in the early part of the flames. Total incident heat fluxes ranged from 50 to 250 kW m^{-2} and were largely radiative because of the strong soot emissions combined with lower gas temperatures and velocities.

Peak heat fluxes are of value in determining the response of stressed components, for example, the 'dry' wall of a fire impinged pressure vessel or a critical steel structural member. When the high flux area is relatively small, however, flame fluctuations due to wind or other factors may over a few tens of seconds reduce the actual average flux at a specific location.

Prediction of fire impinged vessel or structure response also requires knowledge of the total thermal input. For this purpose the spatial average flux over the flame engulfed area is appropriate. These values are summarised in Figure 3 and are computed from the measured flux distributions on the basis that areas exposed to fluxes greater 50 kW m^{-2} are flame impinged. The maximum average flux in the sonic natural gas

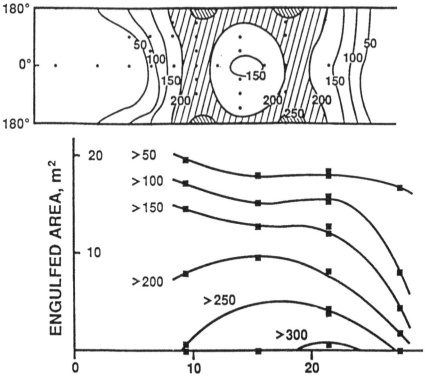

Figure 2. Heat flux distributions on 900mm dia. pipe in sonic natural gas flame. (Top figure is spatial distribution on pipe at 9m).

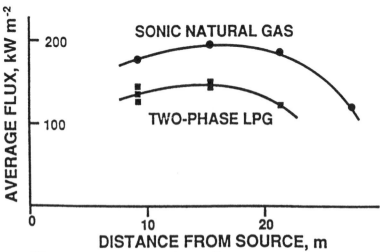

Figure 3. Average heat fluxes over jet flame impinged area.

flames studied was approximately 200 kW m^{-2} and that in the two-phase propane flames was 150 kW m^{-2}. The latter value is not substantially greater than fluxes in pool fires. The average fluxes are also less than previously assumed. Attention must be given to the effects of fuel, radiant flux distribution and scale when applying these measurements to prediction of other flames.

4. CONCLUSIONS

An extensive experimental programme has been conducted that has produced comprehensive information on the properties of full-scale natural gas and two-phase propane jet fires. The findings resolve uncertainties associated with jet fire hazards and provide a sound basis for the better understanding, prediction and reduction of jet fire hazard consequences.

New data has been obtained on the behaviour of jet fires from horizontal discharges. There are distinct differences in flame shape, buoyancy and radiative properties between jet fires from two-phase propane releases and those from high pressure natural gas discharges.

Impingement and engulfment on realistic sized objects is substantial and jet fire thermal impact is not localised.

The heat flux distributions over impinged objects are complex and depend both on flame type and position in the flame. Maximum fluxes are not necessarily at the stagnation point.

Heat fluxes in the sonic natural gas flames studied were 50 to 300 kW m^{-2}. The areas of maximum flux were small and localised within the flames. Average heat fluxes over the total flame impinged areas were approximately 200 kW m^{2}

Heat fluxes in the propane flames were 50 to 250 kW m^{-2}, with average heat fluxes over the total flame impinged area of approximately 150 kW m^{-2}.

The measured heat fluxes considerably reduce previous uncertainties in these quantities and the values are also lower than previously assumed.

5. REFERENCES

1. Cowley, L.T. and Tam, V.T. Pressurised LPG releases - the Isle of Grain full scale experiments. Gastech 88, 13th International LNG/LPG Conference & Exhibition, Kuala Lumpur, October, 1988. Publ. by GASTECH Ltd, Herts, UK., Section 4.3.

2. Cowley, L.T. The thermal impact of LPG (torch) fires. LPGITA Seminar on Fire Protection of LPG Storage Vessels, Solihull, October 1989. Publ: LPGITA, Reigate UK.

3. Chamberlain, G.A. Developments in design methods for predicting thermal radiation from flares, Chem. Eng. Res. Design, 56, 299-309 (1987).

4. Cook, D.K., Fairweather, M., Hammonds, J. and Hughes, D.J. Size and radiative characteristics of natural gas flares, Chem. Eng. Res. Des., 65, 310 (1987).

THE ASSESSMENT OF SAFETY MANAGEMENT SYSTEMS FOR EFFECTIVE LOSS PREVENTION

Barry Whittingham
Principal Engineer
Electrowatt
Consulting Engineers and Scientists
54 Queens Road
Aberdeen

INTRODUCTION

A major concern of the UK Offshore Industry in the wake of the Piper Alpha disaster has been the subject of the Management of Safety. The Cullen report (Ref. 1) has placed great emphasis on the need for the owners of offshore installations to be able to demonstrate that the facilities for which they are responsible are managed safely. The lessons to be learned from the accident to Piper Alpha are drawn out in various parts of the Cullen report, but in particular in Chapter 21 of the report, and notably in Section 21.56. Cullen requires that a demonstration of an adequate Safety Management System should form a leading part of the Safety Case for an offshore installation and should:

- set out the safety objectives
- describe the system by which these objectives are to be achieved
- define the performance standards which are to be met
- state the means by which adherence to these standards is to be monitored

Cullen goes on to list some of the areas of operation which will need to be addressed if safety is to be assessed including organisational structure, management personnel standards, training, procedures for all important operations, monitoring and auditing, accident and incident reporting and investigation to name but a few. A clear duty is placed upon management not only to ensure that proper systems and procedures are in place, but in fact that they are being used on a day-to-day basis.

This paper describes an approach to the assessment of the Safety Management of an existing installation which is currently being applied on behalf of a number of UK North Sea Operators. The paper defines the rationale of Safety Management including the problems which face managers in trying to successfully control the safety of operations which are being carried out, not only in a hostile and unforgiving environment, but which are taking place many hundreds of miles from where management are themselves located. The approach which is described utilises a simple but systematic structure within which the many disparate elements of operational control can be fitted thus enabling each element to be continuously monitored.

PROBLEMS OF SAFETY MANAGEMENT

The rationale underlying the management of safety within an offshore operating company is quite simple in that the senior management of the company is legally accountable for the safe performance of its operation. Whilst the offshore installation is the property of the organisation, senior managers are charged with its care, custody and control and are held accountable because of the discretion they are given, as part of their day-to-day responsibility, concerning the deployment of the company's resources. This includes the way in which the facility is managed in terms of design, commissioning, operation and maintenance as well as the ways in which change and innovation is accomplished.

However in most organisations, senior managers will not make technical decisions concerning aspects of offshore operation on a day-to-day basis, instead, they will establish safety management systems within which activities are undertaken and controlled. They will select and empower staff to operate the installation on their behalf and to do so within a safe operational envelope. Safety Management Systems will be necessary to provide assurance that activities offshore are in fact being carried out safely and in the manner intended.

Safety Management, as with all areas of successful business management, should always be encompassed and controlled by the general principles of Total Quality Management applied across the whole of the company's operations. However, experience reveals that unique difficulties arise in the area of the management of safety so that it is appropriate for this important area to receive special attention to identify and resolve these difficulties.

The difficulties in achieving full assurance that Safety Management Systems are effective arises from the subtle consequences of placing the control of complex technology within a hierarchical organisational structure designed to serve multiple interests. The employees of an organisation, whilst serving their own interests, must, if the venture is to survive, also serve the interests of superiors, share holders, customers and regulators. Each interest served places specific, and sometimes conflicting, constraints upon employees of the organisation. Although formal systems may be instituted to ensure that the important policy objectives are met, it is not uncommon for informal structures to become established in order to help resolve the inevitable conflicts which may arise. Such informal or ad-hoc structures can, unless detected in a timely manner, erode safety standards without the knowledge of senior management. It is important therefore that the safety management systems provide an in built element of feedback as a check that the formal system is in fact being followed and is not being abused.

CURRENT METHODS FOR ASSESSING SAFETY PERFORMANCE

It is well known that plants which are virtually identical in terms of design and operation can have widely divergent standards of performance in safety and productivity. It has been traditional for management to attribute these differences in performance to inherent human variability. However, this is no longer acceptable and all plants which have substantial risk potential are expected to meet the same high standards. The differences, where they exist, must be related to measurable and remediable influencing factors if standards are to be improved. Attempts have been made to solve this problem by developing easily useable indicators of safety performance but such attempts have so far been unsuccessful and have not been widely adopted at least in the UK.

Currently, several methods are available for the assessment of safety performance of an hazardous installation and include

- ISRS (International Safety Rating System)
- MORT (Management Oversight Risk Trees)
- MANAGER
- STAHR
- etc.

It is believed that many of the current approaches have limitations which severely restrict their application and usefulness. ISRS, for example, is principally useful as a points audit system for rating different installations in a comparative way. It does not always provide a complete and balanced overview of safety performance because of the possibility of biased pre-selection of the systems to be audited. On the other hand, MORT has the potential for providing a rigorous assessment of Safety Management of a given installation but in practice is a rather impenetrable and difficult system to put into use.

The methods listed above are indicative of the variety of methods which might be employed to assess the effectiveness of the management of safety. However, these methods on the whole tend to be reactive in nature in that they audit an existing system for faults or incipient problems which are then corrected. Total Quality Management principles suggest that quality is a matter of achieving excellence in all aspects of operation at all times rather than responding to poor quality in a reactive way. An alternative approach to that taken by most of these methods is one which is essentially pro-active rather than reactive in nature. The approach described below organises an existing or a new Safety Management System within a formalised structure which enables excellence to be achieved at all times thus preventing the incidence of latent and incipient fault conditions which, if undiscovered, will eventually result in accidents.

A STRUCTURE FOR SAFETY MANAGEMENT

Safety management operates within the framework of general management whose primary function is:

- to define objectives
- to formulate policies to implement these objective
- to implement the policies by mobilisation of resources

This is congruous with the specific rationale of safety management outlined above and will extend into all aspects of operations within an organisation. The practical implementation of safety management, within the general framework will take place at three hierarchical levels:

Level 1: Statement of **Company Safety Policy** which translates safety objectives into the general policies of senior management.

Level 2: Organisation of the form of **Company Standards**. Such standards will provide a statement of the required performance achievement appropriate to particular operational functions. They will translate policies into intentions and define authority and responsibility of middle management.

Level 3:   Arrangements in the form of **Company Procedures**. These will provide the detailed method of performing specific tasks and will effectively translate the intentions of the Standards into actions at operational level.

The above framework will provide a tangible and auditable means by which the Health and Safety Policies, which companies are legally bound to provide in the form of a statement under the Health and Safety at Work, etc, Act 1974, can be translated into working practice. Most companies will therefore have issued at some time a statement of Policy but many will be unable to demonstrate that specific policies are being implemented by appropriate procedures which exist within the company at operational level. The institution of Company Standards as defined above is intended to provide this essential mapping function between Policies and Procedures. Not only will Company Standards enable senior managers to be aware of which procedures exist at offshore operational level, but they will also define the responsibilities of staff for issue and control of procedures. Immediately therefore, the existence of Company Standards of the type described, will place senior management in the powerful position of being able to exert some form of control over the safe operation of the investment for which they are ultimately responsible.

However, as will be described below, the mere knowledge of the existence and location of procedures is insufficient to provide the assurance required by senior management as recent major accidents have made abundantly clear.

SAFETY MANAGEMENT ASSURANCE

It is important that any framework for reviewing Safety Management provides assurance to management that safe working is being implemented in practice. The emphasis here is always upon conformance i.e. that human behaviour conforms to established procedures. It is essential therefore to ensure, not only that policies, standards and procedures to control safety are in place, but also that they are being used. In reality, the method for assuring conformance will be a dynamic system for independent feedback on performance standards for the appropriate level of management to respond to. The fact that the feedback loop is dynamic ensures that the organisation has sufficient flexibility to respond adequately to change. The continuity of the loop, whilst ensuring that management maintains a state of awareness and vigilance about operational safety, also has the important function of creating an awareness of the Company's safety objectives, and how these are to be achieved, at operational level.

The loop comprises three main components :-

-       feedback of performance indicators
-       decision on required actions
-       action to implement decisions

and is identical to other simple operational control systems which are to be found at all levels of an organisation.

The main criteria for the design of the systems to assure compliance is the need to achieve a balance between two extremes:-

a)   Reliance on rapid "ad-hoc" judgements at too low a level of management, resulting in under control and loss of assurance for the higher levels.

b)   Over control at too high a level of management by excessive data collection and deliberation (or "paralysis by analysis").

Some general guidelines to control are:

-       Slight over-control is better than loss of control
-       It is easier to remove or ease controls then install them later
-       More effective results are obtained from short control cycles, i.e. rapid feedback loops
-       Controls should be delegated as far down the management hierarchy as possible to achieve a short control cycle
-       Risk decisions should be delegated upwards to the appropriate level of management responsible for that risk level
-       No control at all is only acceptable when the feedback/decision/action loop protocols are such that they are safely managed at operator level.

Just as safety is congruent with all other operational goals, so the balance of control of safety will produce benefits in productivity through more effective use of resources and enhanced staff morale and minimise waste and efficiency by eliminating unnecessary deviations.

CRITERIA FOR EFFECTIVE SAFETY MANAGEMENT

The reports of inquiry of many major recent industrial and other accidents have highlighted deficiencies in Safety Management as the Root Cause of the accident (Ref. 2). A review of these reports, in an attempt to draw out the main lessons to be learned for Safety Management has led to the definition of three main criteria for reviewing the Safety Management, Systems for a given operation. These are :-

1.   **Completeness** - to ensure that safety management systems exist.

This involves a comprehensive review of all the main ares of operations with respect to the provision of Safety Management systems for each area in order to identify omissions and assure ultimate completeness. Each main function within each of the operational areas (i.e. production, maintenance, etc) will then be studied to identify:

-       relevant safety management system provided
-       method of implementation
-       management directly responsible

2.   **Adequacy** - that the systems which exist are fit for the purpose

This is a qualitative assessment of the adequacy of the safety management systems:

-       to achieve safe and efficient operation
-       to assure senior management of safe and efficient operation

It will assess adequacy by considering issues such as:-

a)   effectiveness in translating policies into practical action
b)   quality of communication of safety management policies
c)   quality of feedback from operational to management level
d)   management responses to practical safety issues
e)   delineation of responsibilities
f)   level of direct and overview responsibilities in the organisation

This list is indicative only of the issues to be studied and is not intended to be exhaustive.

3.   **Validation** - that the systems are being used properly

The emphasis here is on the important issue of providing assurance that the Safety Management systems provided are being routinely implemented in practice and as intended and will include:

-      validation of conformance using feedback, decision and action control loops.
-      overview responsibility for validation.

The objective is to identify the systems that are provided to ensure conformance to the operational procedures, and not to assess the adequacy of such systems as described above.

If it can be demonstrated that the three criteria of completeness, adequacy and validation are being met for a given installation, then it is believed that a Safety Case can be made for that installation regarding the effectiveness of Safety Management.

CONCLUSIONS

An effective method of assessing the effective management of safety of an installation must start from the premise that the management of the company must have the assurance of full control of all safety critical activities which are taking place. Although most companies issue Health and Safety Policies, not all companies, are able to demonstrate the precise manner in which these stated policies are being implemented in practice, i.e. the management of a company may be aware that activities carried out at operational level are subject to set procedures but cannot necessarily demonstrate that this is the case in practice.

The management of some companies may not be able to readily locate the procedures that are supposed to be used within the company, because no system of recording responsibilities for issue and control of procedures exists. The first requirement therefore of any effective system of Safety Management is to be able to establish connectivity between Company Health and Safety Policies and Operating, Safety and Emergency Procedures. The method of achieving this connectivity is by means of Company Standards as described above.

The Company Standard translates the intentions of the Company Safety Policy into practical implementation at operational level and provides the essential demonstration that this is being done. An assessment of Safety Management within a company will therefore set out to establish the essential element of connectivity between policies and procedures and if this does not exist then the company must make proper provision for it.

One of the more interesting features of the review method which has been described is the possibility of developing a computerised data base which is able to map the line responsibilities for safety within an organisation to actual operational procedures and methods. The objective is to develop a concept which is analogous to that of a "Living FSA" which is able to take account of equipment modifications at a future time and very quickly and easily predict the impact of these changes on the safety of the installation. In the same way it is believed to be possible to accurately monitor the impact of organisational changes on the safety of a given installation. Such a development will of course have major benefits in day-to-day operations of being able to document the control, issue, use and modification of procedures at operational level. Many more benefits of such an ordered and rigorously controlled approach can easily be foreseen.

The basic requirements have already been set out above for carrying out an assessment of safety management within an organisation which would allow such a computerised "mapping" approach to safety management responsibilities through an organisation to be implemented. However, no such system is currently being implemented within industry due to commercial restraints on the part of companies. It is appropriate therefore that research into the development of computerised systems of Safety Management should take place in the future.

REFERENCES

1. The Public Inquiry into the Piper Alpha Disaster. The Hon. Lord Cullen. London, HMSO, November 1990.

2. Whittingham, R.B., The Application of Root Cause Analysis to Incident Investigation to Reduce the Frequency of Major Accidents. SARSS '90 Conference Proceedings, The Safety and Reliability Society. September 1990.

SESSION C:DETECTION AND CONTROL

THE SELECTION AND PLACEMENT OF FLAME DETECTORS
FOR MAXIMUM AVAILABILITY OF THE DETECTION SYSTEM

Garth Watkins BSc MBA

Detector Electronics (UK) Ltd.

SUMMARY

In the wake of the Piper Alpha disaster all aspects of offshore safety have been closely examined and changes will follow in the design, use, placement and maintenance of safety devices. Flame detection continues to play a vital role in ensuring the safe operation of onset and offshore installations.

Even prior to July 1988 many operating companies in the Offshore Industry had reviewed the performance of their safety systems and in particular the fire protection. One of the findings was that the time available for detection was far lower than expected and there were many contributing factors.

The paper will examine these reasons and illustrate how products employing newer technologies and CAD systems to locate detectors can significantly enhance the reliability of the safety system.

INTRODUCTION

For some time North Sea operators have been concerned about the effectiveness of their fire protection systems and following the Piper Alpha disaster the subject has received greater attention. The Cullen report and revised draft of SI611 are only part of the review that is taking place.

Fire protection begins with detection and there are many ways of detecting the presence of combustion. However, the environment of application often confines the means of detection to a small choice and offshore production platforms and drilling rigs are good examples of hazardous areas which are either open to the environment or ventilated such that smoke and even heat detection are ineffective. As a result flame radiation detectors have been widely used in the petrochemical industry.

Technical evolution resulted in the introduction of solar blind ultra violet detectors twenty years ago and more recently in infra-red detectors. These newer devices offer advantages over the UV detectors but they too are not perfect. A system employing any form of flame radiation detection must account for the strengths and limitations of the detectors, as must the operation and maintenance procedures to ensure the integrity of the system as a whole.

HISTORY

Very early flame detectors were designed to sense infra-red radiation in the 1-3um wavelengths. These devices were sensitive to fire but also very sensitive to background radiation which meant that they false alarmed easily and were quite useless in areas where daylight, let alone sunlight, was present (Figure 1).

About 20 years ago, refinements to the Geiger-Muller tube technology resulted in sensors that were only sensitive to photo emission in the very short wavelengths, in the region where the ultra violet radiation from the sun is absorbed by the ozone in the earth's atmosphere. Having achieved solar blindness, electronic signal processing enabled manufacturers to produce UV detectors that were sensitive to flame radiation over reasonable distances, like 15 to 20m and UV flame detectors soon became the industry norm.

Infra-red radiation continued to interest the industry and advancing technology produced new optics and signal processing techniques which enabled the sensing of radiation in a narrow band around 4.4um. About 7 years ago the first IR flame detectors came on to the market that were solar blind and these are now becoming a major player in the optical flame detection field.

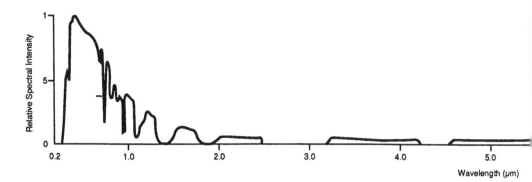

Figure 1 Solar radiation measured at surface of earth

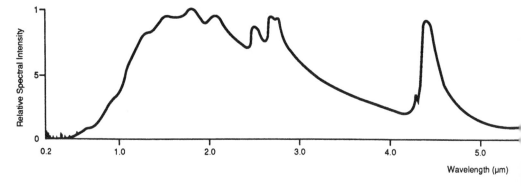

Figure 2 Radiation from hydro carbon fire

TECHNOLOGIES

In order to engineer a flame detector very detailed examinations of radiation spectra, not only of fire but also many false alarm sources, are undertaken. Starting with the fire, one needs to define the fuel, size of fire and distance over which the detector is to detect it. The most common fuel is the family of hydrocarbons and a typical radiation spectrum is shown in Figure 2. This at least shows that radiation is present across the whole spectrum we wish to consider, albeit at differing intensities.

The examination then turns to false alarm sources and the most difficult by far is the sun. Fortunately the atmosphere absorbs radiation at various wavelengths; ozone at the UV end and CO_2 and water at the IR end of the spectrum, as shown in Figure 1. If other sources are added to the list, it soon becomes apparent that the choice of suitable detection wavelengths is very limited. If we overlay the radiation spectra of various false alarm sources it is apparent that UV at less than 0.25um and IR at 4.4um are the best choices.

As mentioned before, UV radiation is best sensed with a Geiger-Muller tube but it is necessary to significantly enhance the sensitivity because the signal intensity is very small (Figure 2). This can be achieved by using relatively large areas of electrodes of very pure metal and by this means UV detectors are very sensitive to small amounts of radiation and so can sense a small $0.1m^2$ petrol fire at 20m.

At the infra-red end of the spectrum the task looks easier since the signal is large and, providing the sensor "window" is limited to the wavelengths around 4.4um, a false alarm free detector should be possible. The sensors are mostly pyro-electric or thermopile devices used with patented narrow band filters concentrated on the 4.4um band and this combination makes solar blind detectors a reality.

However, there is the question of false alarms from hot black body radiation, of which the incandescent bulb is an example. If the temperature of a black body is raised to 100°C (with 100% of the field of view filled with the radiation) the IR sensor will detect it. Thus, hot black bodies become a potential false alarm source and the principle technique used to reject this source is to sense a flicker as well, since hydrocarbon fires flicker at characteristic frequencies in the range 1 to 10Hz.

Not all fires radiate at 4.4um, which is the characteristic wavelength associated with the vibration of carbon dioxide molecules. The formation of CO_2 in the flame results in emission at this particular wavelength and if the fire is not carbonaceous, this emission will be missing. This is true, for example, in hydrogen, sulphur and metal fires and IR detectors are not suitable where these fuels are the risk. Equally, not all fires radiate at the short UV wavelengths and glycol is an example of a fuel which emits very little UV and is therefore difficult to detect when burning, using a UV detector.

In addition to false alarm sources, consideration must be given to the presence of substances which may absorb the radiation at wavelengths the detectors are designed to sense. Ultra violet wavelengths are absorbed by oil and for this reason detectors should be equipped with an optical integrity checking system to alert a user if a detector window becomes contaminated to the point that its sensitivity is seriously impaired. Such a checking system can also be used to check the electronic signal processing and system circuitry to ensure that the integrity of the detection system is maintained. Equally, UV detectors are not suitable if UV absorbing solvents such as toluene or acetone are likely to be present in the atmosphere (Ref 5).

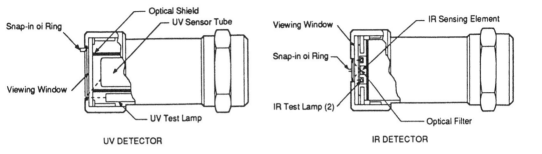

Figure 3   Optical Integrity Test

Infra-red detectors are similarly affected by absorbing
substances and the most serious is water, ice and snow.   Water
in all its forms absorbs infra-red radiation (Ref 1) and will
adversely affect an IR detector.   Grime and dirt deposits are a
bigger problem for UV detectors, although there is a level of
contamination that will affect an IR detector too and neither
will perform well when the windows are covered with mud or
indeed a coat of paint.   Thus the need for checking that the
window will transmit radiation emitted from a fire is essential
and if this mechanism is built into the detector and operates
automatically, so much the better.

CURRENT TRENDS IN DETECTOR SELECTION

Whilst ultra violet flame detectors are still the most widely used detectors in the world, single frequency infra-red detectors are being increasingly used.

Some of the limitations of both forms of detection were discussed above and there are others which are worthy of consideration.

One of the most intense sources of UV radiation is present in the electric arc found in arc welding and UV detectors are, not surprisingly, easily alarmed by this source. As a result, part of the procedures when doing such hot work is to isolate the UV detectors in the affected area for the duration and post a fire watch, since arc welding is also a very obvious ignition source. Another false alarm source is nuclear radiation such as gamma or X-rays, which may be used in non destructive testing. These rays cause the Geiger-Muller tube of a UV detector to conduct, thus simulating photo emission and setting the detector into alarm. Again, it is necessary to isolate the UV detectors in the area where NDT is to be undertaken or employ a specially designed system known as nuclear surveillance, which prevents false alarms from nuclear radiation.

Neither arc welding nor nuclear radiation cause infra-red detectors to alarm in normal use. Of course, if the arc is close enough to the detector, it will sense the small amount of IR radiated, but this is not generally a problem at distances greater than 1m for single frequency IR detectors.

Oil contamination is a problem for UV detectors in areas where there is oil mist present. However, the optical integrity (Oi) check must be used as a warning that the detector needs cleaning. Unfortunately there are examples of installation where these warnings have been abused. On one platform the detectors were covered with plastic bags, which had the effect of eliminating Oi faults but it also rendered the detectors useless, as not only would the oil on the bag absorb the UV but so too would the bag itself.

As UV detectors employ pure optics, it follows that all the optical surfaces must be clean, including the reflector ring used in the Oi system. This can be difficult in areas where airborne contaminants are present (not only oil but dust). The introduction of an internal reflection system of Oi significantly simplified the cleaning process and this has now been further simplified by the introduction of a sensor module which does not require a special lens to conduct the internal reflection test and can therefore be used in standard detectors. The limitations of UV detectors discussed above have resulted in many fires not being detected by the automatic detection system. One operator found that this was as high as 60% of fires, largely because the system was isolated during arc welding and NDT.

The introduction of solar blind infra-red detectors has given the industry a new opportunity to deploy flame detectors and use them during times of greatest risk, i.e. when hot work is taking place. The single greatest limitation of these detectors is that they are blinded by ice on the lens. Water is less of a problem since the windows are hydrophobic. However, water will adhere to other contaminants and tests have shown that in extreme conditions it is possible to get a thin film of water across the lens of a detector. Absorption of the radiation by rain or snow is a fact, but other tests have shown that the effect is to reduce the detector sensitivity by about 20% and if this is taken into account in the system design, the performance of the system will be adequate even during a downpour. The use of an optical integrity test will reassure an operator of the system availability.

A further advantage of infra-red detectors is their ability to detect very smokey fires such as those of diesel. The black smoke absorbs UV radiation and if the detectors are sufficiently distant from such a fire, it is possible for the UV to be attenuated so that a UV detector will not detect the fire. Infra-red radiation is not absorbed to the same degree and the fire is therefore still detectable by IR detectors.

SYSTEM DESIGN

With the imminent introduction of Formal Safety Assessments in the UKCS, the deployment of flame detectors becomes all the more important as it will no longer be sufficient to scatter detectors around without carefully checking that the system is "fit for purpose". This means that the risk will need more careful defining so that the logic of the detection system can be commensurate with this risk. One operator has developed a grading system which defines areas of high, medium and low risk (Ref 3). In high risk (Grade A) areas, only small fires (>10kW radiant heat output) can be tolerated for short periods and so the detection and protection system must react quickly to contain the fire and limit the damage. The medium risk Grade B areas can tolerate fires of up to 50kW radiant heat output before a control action is taken and low risk (Grade C) areas do not require a control action until the fire size exceeds 250kW radiant heat output. Thus the fire can range from a small $0.25m^2$ jet fire to a large $12.5m^2$ pool fire before control actions are initiated, depending on the risk.

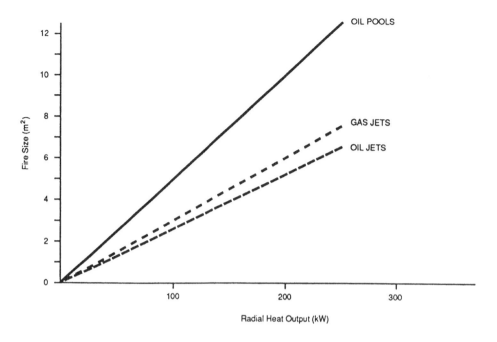

Figure 4   Approximate Correlation between
Fire Size and Heat Output

Having graded every area to be protected, and in particular
identified points of highest risk in the overall design, it is
possible to lay out a detection system in such a way that it
fits the parameters of the design specification including
thresholds of fire size to be detected and the voting logic to
be employed to initiate control actions.

SYSTEM MAINTENANCE

The performance of a system which has been carefully designed
to meet all aspects of the specification is only as good as its
maintenance. Since radiation flame detectors need to "see" the
flame, any obscuration will effect the system performance.
Process areas on production platforms are never fixed in their
design and it is therefore probable that a detector will need
to be moved during the life of the system. A pipe run may pass
directly in front of a detector or a new piece of process plant
be installed in an area which will affect the system's ability
to detect fires and vote as intended. The most frequent
obscurer of flame detectors is scaffolding and routine checks
are needed to ensure that such obstacles are not blinding a
detector. The worst example of this was a permanent platform
installed to enable an instrument technician to reach a UV
detector which was high up in a process area, to clean and
maintain it. The support for the platform was directly in the
line of sight of the detector.

One of the great benefits of radiation flame detectors is that
they monitor a volume. The detector can sense throughout its
cone of view and this is important to remember when installing
and maintaining detectors. It is wise to limit the cone of
view to the risk being monitored and not include backgrounds
which may contain false alarm sources in the cone.

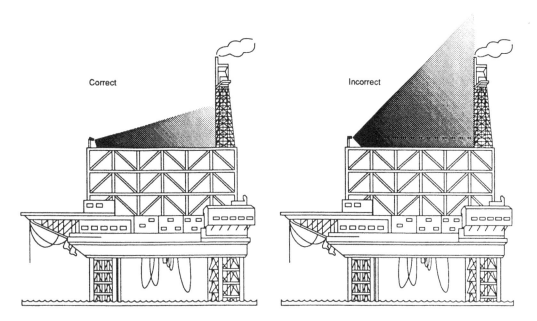

Figure 5 Correct Detector Orientation

Figure 5   Correct Detector Orientation

Having carefully set up a detector to monitor an area, its angle of view must be audited from time to time to ensure that some obscuration has not entered its view and that it is still pointing in the correct direction.

Periodic checking of detectors has been mentioned and a built-in Optical Integrity check is a convenient and reliable means of achieving this. It may also be desirable to check a detector performance with a test torch. However, these must be used with care because they are not generally calibrated to represent a fire. One operator had assumed they were and when a UV detector indicated an Oi fault, he had checked it using a UV test torch, which brought the detector into alarm and he therefore assumed that all was well and that the Oi fault indication was wrong. The UV emitters in test torches are high intensity emitters and are only useful to check the line of sight and function of a detector.

Similarly, infra-red detectors require a large amount of energy to cause them to alarm and most test torches need to be relatively close (< 2m) to a detector to check its function.

Nevertheless, such detector testing is worthwhile and indeed essential if detectors are not fitted with a built in optical integrity checking feature.

CONCLUSION

The detection industry continues to research new and better ways of detecting fire and ignoring false alarm sources. There is still no perfect fire detector, although current generation IR detectors have many advantages over UV detectors. The next generation will be more sensitive, less false alarm prone and more difficult to test. The perfect detector will of course need a real fire to test it - a prospect that needs careful consideration.

Advances in fire detection techniques, particularly resulting from the introduction of less false alarm prone infra-red detectors, will help system specifiers and designers to install more complex fire protection in hazardous areas, which more accurately meet the need of the risk. These newer detectors can remain on-line even when exposed to arc welding and NDT radiation, thus significantly increasing the availability of the detection system. It is anticipated that the proportion of fires detected by the automatic system will increase significantly and this should result in a reduction in losses.

ACKNOWLEDGEMENTS

The author wishes to thank the following for their valuable contribution to the content of the paper:

R.J.C. Bonn, B.P. Petroleum Development, Fire Engineering HSEQA

I. Davidson, Micropack

J. Vogel, Detector Electronics Corp.

D.N. Ball, Kidde-Graviner

REFERENCES

1. Alienikov V.S., Belyaev V.P., Devyatkor N.B., Mamedly, L.D., Masychev V.I. and Sysoev V.K., 'CO Laser Applications in Surgery'. Optics and Laser Technology, October 1984.

2. Ball, D.N. 'Flame Radiation Detectors for Fast Fires and Explosions'. Technical Programme - 'Electronics in Oil and Gas', London 1985.

3. Bonn, R.J.C. 'Practice for the Specification of Fire Detection Systems for Offshore Installations'. BPPD Fire Engineering HSEQA. Draft Issue June 1990.

4. Bonn, R.J.C., Barson, J. 'Flame Detector Evaluation'. BPPD R & D Report ER 2884.88, March 1988.

5. Detector Electronics - 'Fire Detection Selection Guide' 92-1002-01 May 1983.

6. Larsen, T.E. 'Signatures et Forgeries - A discussion of Fire Detection using UV and IR Portions of the Spectrum'. Detector Electronics Seminar, London 1983.

7. Skippon, S.M. 'Suitability of Flame Detectors for Offshore Applications'. Offshore Contingency Planning Conference, Aberdeen, May 1989.

Use of ESD valves in Fire Safety Engineering based on Safety Assessments

by
Odd Thomassen, Statoil
and
Jan Erik Vinnem, Safetec Analysis Ltd.

1   INTRODUCTION

Much attention has since the summer of 1988 been focused on fire loads on platforms, and the prevention of such loads leading to escalation of an accidental event, into something which may grow beyond control.

Emergency shut down valves on risers have been given the primary attention, for new as well as existing installations, However, also ESD valves in the process area should be considered carefully.   Most fires will start here, and it is by the safety systems in the process area that the events well have to be controlled.

The basic requirements for safety protection of the process systems are spelled out in the API RP 14C (Ref. 1). Process safety systems as well as emergency safety systems requirements are presented in this standard.   The Norwegian Petroleum Directorate (NPD) has required (Ref. 2) isolation valves, in order to limit the possible fire potential. Some uncertainty has thus been created over the years, as to whether Emergency Shut Down (ESD) valves or Process Shut Down (PSD) valves were required.

At the same time, risk assessments are being used more and more extensively. One of the primary uses of risk assessments is a systematic basis for definition of what should be Design Accidental Events and associated design loads.

This paper presents a systematic procedure for a logical definition of the applicable valve type, either as ESD or PSD. The approach is based on use of risk assessments and design accidental loads, and is intended to ensure that the installation of shut down valves is consistent with the requirements for fire protection in the process area, in order to maintain a high safety level. Blast loads are also important for the protection of ESD valves. These loads are not considered in this paper.

The difference between a PSD and an ESD valve may not be excessive, when valves are specified for new installations. It is often customary to use valves for the PSD service which also will satisfy the fire load integrity requirements for ESD service. The differences may therefore be limited to the control side.

The difference may be much more significant for existing platforms. Retrofitting ESD valves for existing PSD valves may be expensive as well as complicated, and should only be implemented where logically required. Still, safety can not be compromised, and a systematic approach will therefore be advantageous.

The paper will present the fundamental requirements for shut down valves for ESD as well as PSD service, as a basis for discussing the proposed procedure.   The approach adopted has recently been used to define needs for upgrading of valves on an existing installation.   The implications of the procedure for a practical case is therefore illustrated on this basis.

2 DEFINITIONS

This section presents definitions that are essential for the approach developed for valve classification. Requirements for different valves in the process system are presented in Sections 3 and 4.

2.1 Design Accidental Events

Design Accidental Events are used as defined by NPD in their Guidelines for safety evaluation for platform conceptual design. (Ref. 3)

A Design Accidental Event gives the basis for requirements for the main safety functions as defined by NPD (Escape Ways, Shelter Area, Support Structure, Ref.3)

The Design Accidental Events are also used to define the Design Accidental Fire Loads.

2.2 Design Accidental Fire Load

The Design Accidental Fire Loads are expressed as heat flux (kW/m^2) as well as duration (minutes/hours) which are used as the design basis for a specific fire area.

Different design fire loads may result from pool fires as well as gas fires within one fire area

2.3 Fire Area

A fire area is often limited by passive fire walls. However, also distance may be used to segregate fire areas without any physical restrictions.

3 API REQUIREMENTS

The basic requirements for process safety systems are found in API standards, as outlined below. These requirements have been established years ago, and are very widely used.

This section presents the main safety devices in the process area according to API standards.

3.1 Shut Down Valve (PSD)

An automatically operated normally closed valve, which according API RP 14 C (Ref. 1) is installed in order to isolate a process station.

A PSD valve shall be activated to a safe position, which often is the closed position.

3.2 Pressure Safety Valve (PSV)

Valves which according to API RP 14 C (Ref. 1), and API RP 520 (Ref. 4) shall open for given overpressure upstream of the valve. Pressure safety valves are normally spring or pilot operated.

3.3 Blow Down Valve (BDV)

Valves, which according to API RP 521 (Ref. 5) shall be a "fail safe open" valve, for pressure relief of a process component.

3.4 Process Control Valve

Valve used for control of hydrocarbon processes, such as flow control or pressure control valves. These valves are operated by the process control system.

3.5 Check Valve

Valves, which according to API RP 14C (Ref. 1) should be installed in order to protect against backflow of gas or liquid of significant volumes in the case of an accidental leak.

4 NPD REQUIREMENTS

NPD requirements for safety systems in the process area are first of all according to API as defined in Section 3. In addition NPD requires Emergency Shut Down valves (ESD) as isolating valves in order to limit the maximum fire potential.

Emergency Shut Down valves are automatically operated shut-down valves installed in order to isolate hydrocarbon volumes according to design fire loads.

The ESD valve shall first of all satisfy requirements for shut-down valves (PSD). In addition the following requirements apply to the ESD valves:

- Satisfy fire testing requirements according to BS 5146 (Ref.6).

- The position of the valve shall be indicated in the control room in addition to a local indicator. The information given in the control room is as a minimum end positions as well as valve operation.

- The ESD valve shall fail to safe position, which often implies a fail closed valve.

- The valve shall close down production activities. Activation of the valve shall be either closure or opening, but can not be both functions.

- In the case that hydrolic or pneumatic energy is required to operate an ESD valve the accumulator will need to be located close to the valve. The capacity of the accumulator shall be sufficient for 3 valve operations.

Reset of an ESD valve following an emergency shut-down activation shall only be possible from the local panel, following permission from the control centre. Reset shall normally only be allowed with the main control system, not with pressure from the accumulator. This is according to the principle that production requires availability of the main control system.

For ESD valves in the wellhead area or on the wellhead certain specific requirements apply which are not covered here.

An ESD valve may also be used for PSD service if a separate process control system is installed. This may however, lead to escessive seat wear, which has to be taken into consideration.

The following are additional requirements for and ESD valve in order that it shall function as intended with respect to isolation in leak and fire cases:

- Operation of the valve must be possible with full flow in the line

- Operation of the valve must also be possible under fire load impact

- The closure time for the valve must comply with premises used in the assessment of the design fire loads.

5 FIRE LOAD CRITERIA

5.1 Fire Integrity

All fire partitions (either as physical partition or as distance) between two separate fire areas will have to be designed to maintain its integrity under the design fire loads for the areas.

5.2 Assessment of Design Fire Load

This section spells out general premises for assessment of design fire loads. The actual assessment of such loads is given separately for liquid fires and gas fires in Sections 5.3 and 5.4 below.

A fire load assessment for a closed module will have to reflect limitations to oxygen supply due to capacities and characteristics of a mechanical ventilation system. On the other hand may a potential initiating explosion open up module walls and give additional air supply.

In the case of significant explosion overpressure additional air supply may also be created for semi-enclosed modules. This will also have to be taken into consideration. Explosions may also lead to secondary ruptures of other process systems, and the combined effect of leaks from different systems may then have to be accounted for.

The detailed premises and assumptions which should be used for liquid as well as gas fires are outlined below. These apply to the contents of pressure vessels and associated piping.

5.3 Liquid Fire

The design fire loads in a case of a pool fire should be assessed based upon the following premises:

- The maximum contents of hydrocarbon which can exist within a process section

- The cross sectional area of the leak should be a high value implying that the leaking rate of the hydrocarbon is high. This implies that the duration of the pool fire will be significantly longer than the duration of the leak.

- A realistic assessment should be performed of the area on to which the leak

is spilt. This implies that the position of possible leaks will have to be assessed in relation to obstructions such as drip pans. The capacity of drip pans as well as the location will have to be considered. Possible grated floors will also be taken into account.

- The regression rate (rate of combustion expressed as height of liquid film burning per time and area unit) will have to be realistically assessed according to the relevant type of hydrocarbon liquid.

- The duration should be assessed without consideration of drain systems.

- Possible ventilation shut-down in the case of fire detection should be considered for closed modules.

5.4 Gas Fire

The design accidental loads for gas fires should be assessed based on the following premises:

- The amount of gas leaking will take the volume between isolating valves into account.

- The duration of gas jet fire is strongly dependent on the mass flux, which again is determined by the cross sectional area of the leak. A large hole implies a high mass flux and a short duration. The design case should be a relatively small cross sectional area which gives the significant duration. The realistic leak area must be related to the dimensions used in the area. A typical leak area may be full bore rupture of an instrument connection such as a 3/4" line.

- The calculation of volumes will have to consider the time required to activate pressure relief systems according to available systems and relevant procedures.

- A fire jet may expose equipment in any direction and all systems within a fire area must be considered potentially exposed.

5.5 Fire Areas

A fire area is often enclosed by passive fire partitions, of minimum H-0 design, in order to limit the systems that may be exposed to fire loads. The design fire loads will be the basis for establishing what capacity that the fire partitions need to have.

A fire area may also be segregated by distance alone, without any fire partitions. This implies that the distance must allow the design fire to burn without exposing the surroundings to excessive fire loads.

5.6 Availability Requirements

The availability requirements in this section applies to the need to isolate process sections in the event of a fire, in order to limit the fire loads. The consideration discussed herein applies to the isolation function, and not the process control systems.

Availability of the isolation function implies in the present context that the following requirements must be satisfied:

- The valve must close on demand as intended without failure

- The valve must initially be tight in both directions, and must continue to isolate completely, even if a high pressure gradient across the valve exists.

The selection of PSD or ESD valves is dependent on these two factors, in order to prevent fire escalation. High reliability of the isolating function may be achieved by:

- a PSD valve, if process shut down is initiated as part of the process safety function, and the activation system has a high reliability

- an ESD valve, if the valve or its activation appliances may be subjected to the same fire as the valve shall isolate against

This evaluation assumes that an ESD valve has a higher level of protection against leaks through the valve, in the case of a fire load impinging on the valve or its controls.

6 VALVE CRITERIA

This section presents the criteria for selection of ESD and PSD valves for liquid systems in the process area, for all stages of separation, produced water as well as stabilization of crude oil. Similar criteria are developed for gas systems, but are not presented here.

The starting point for this process is the need for isolation between two pressure vessels, and a comparison of the duration of a potential fire with the duration of the design fire.

The consideration is split into separate cases, according to whether the two pressure vessels are installed in two separate areas, or in the same area.

6.1 Valve Installation with Two Areas

A valve of PSD type is sufficient, in the case of the isolation valve located in a separate fire area. This condition further assumes that a check valve has been installed downstream of the fire partition, upstream of the process unit to be protected.

A PSD valve will as an isolation device has a high reliability, as the valve is not exposed to the fire from which protection is sought.

These principles apply to hydrocarbon liquids as well as produced water.

6.2 Valve Installation with One Area

Two conditions may apply regarding what valves that are needed, in the case that two vessels are located within the same fire area:

- The valve can be of the PSD type, if the design fire corresponds to the contents of both vessels.

- The valve has to be of the ESD type, if the design fire condition is limited to the contents of only one of the vessels.

A PSD valve will not provide sufficiently high reliability of the isolating function, due to the possibility that the valve (or control) is subjected to the same fire as it shall provide protection against.

6.3 Valve Installation relative to Pipelines

A valve between the process systems and pipeline (export or import, to/from other platforms, subsea installations, etc) shall always be of ESD type, due to the normally extensive duration of a fire caused by upstream (in the case of an import line) or downstream (export) leak.

7 IMPLICATIONS

This section presents an illustrative example on the implications of using the principles from Sections 5 and 6 in practice.

Figure 1 shows location of isolation valves between 1. stage separator and downstream pressure vessels, on the liquid side (such as

2. stage separator or produced water tank). The module M2 is the separator area (1. stage), whereas the module M3 is 2. stage separation area. The M4 area is the produced water treatment area.

The following are the alternatives as far as isolation valves are concerned, based on the principles above:

- The valve in position "C" will be a PSD service valve, if fire partition is installed between M2 and M3 areas. This also assumes that there is a check valve upstream of the 2 stage separator, downstream of the fire partition. (Pos. "H").

- The valve in position "C" will be a PSD service valve, even if no fire partition is installed between M2 and M3 areas, if the design fire load for M2 and M3 areas is based on the volume of 1. stage or 2. stage separators.

- The valve in position "C" will be a ESD service valve, if no fire partition is installed between M2 and M3 areas, if the design fire load for M2 and M3 areas is based on either of the volume of 1. stage + 2. stage separators.

As regards the function of valve "C", the following scenarios may develop:

A fire in M2 will not affect equipment in M3 (and vice versa), if there is a fire partition between M2 and M3 areas. This implies that a check valve in M3 will not be affected by a possible fire in M2. The pressure vessel in M3 will accordingly be safely and reliably protected from feeding the fire in M2, by this check valve. The same applies for a fire in M3.

The assessment of design fire load for M2 will have to be based on the maximum fire potential in M2, namely the contents of the pressure vessel in M2. No particular requirements apply to the valve in M2, when the fire is in

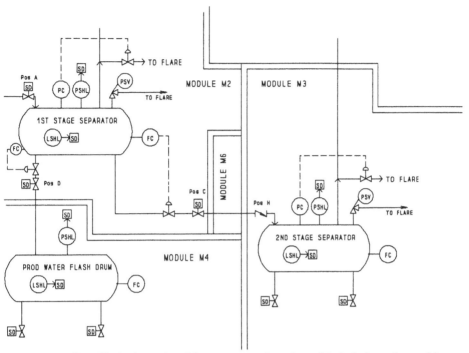

Figure 1   Simplified schematics of 3 process vessels and possible isolation valve positions

the same module. Hence, an ESD valve is not required in the case of a fire partition between the modules.

The design fire load is the decisive factor, when no fire partition is installed between the two modules. No particular requirements apply to the valve between M2 and M3, if the total contents of hydrocarbons in both modules M2 and M3, is the design basis. The valve may hence be of PSD type.

The valve in position "C" needs to be of ESD type, if the design fire load corresponds to the volumest one (the highest) of the pressure vessels in M2 and M3. An ESD valve gives the highest reliability of the closing function, in a fire scenario. This is required to limit the duration of the fire to what corresponds to the

design basis, as the valve may be subjected to the fire loads. Similarly may isolation of the pressure vessel in M2 be required, in the case of fire in M3.

Exactly the same requirements apply to the valve in position "D", between 1. stage separation and produced water treatment tank.

It should be noted that no shut down valve is required in Module M6, if only one isolation valve is installed between to pressure relief points. This implies that the line in M6 is relieved either in M2 or M3.

8 CONCLUSIONS, RECOMMENDATIONS

This paper has presented an approach to optimization of ESD and PSD valves, which-may be an important contribution to enhanced safety, particularly for existing installations.

Process Shut Down (PSD) valves are used for process safety purposes, but may also fulfil isolation purposes in certain circumstances.

What is important, is therefore to distinguish where such service can be allowed, and where it can not be defended.

It should also be noted that many safety engineers would assume that the valves discussed in this paper always should be ESD type valves, irrespective of the conditions and loads that apply. In this way, application of this approach may imply a simplification of the design, and a cost saving measure.

9 REFERENCES

1. Recommended Practice for Analysis, Design, Installation and Testing of Basic Surface Safety Systems on Offshore Production Platforms, API RP 14C, Second edition, January 1978

2. Regulation for production and utility systems, NPD, Stavanger, 1978

3. Retningslinjer for sikkerhetsmessig vurdering av plattformkonsepter, Oljedirektoratet, 1.9.81

4. Recommended Practice for the Design and Installation of Pressure Relieving Systems in Refineries - Parts I and II, API RP 520

5. Guide for Pressure relief and Depressuring Systems, API RP 521

6. Specification for Inspection and test of steel valves for the petroleum, petrochemical and allied industries, BS 5146

MODERN METHODS OF DESIGNING FIRE AND GAS DETECTION SYSTEMS

R J C Bonn (BP Exploration)

SUMMARY

Fire and gas detection should play a crucial role in loss prevention on many sites. CIMAH cases and Formal Safety Assessments usually assume that fire and gas detection systems will reduce risks, yet design is often a matter of "black art"; it is difficult to quantify the parameters involved and there are no guidelines to define required performance or to relate achieved performance to safety requirements.

This paper introduces methods of

- developing a system performance specification based on the hazards expected, with requirements in terms meaningful to systems designers and auditors

- assessing design adequacy and for auditing existing systems.

The methods presented are compatible with HSE CIMAH and Formal Safety Assessment requirements. They have been used for the design and assessment of F&G detection systems on several on- and off-shore petroleum production sites.

NOMENCLATURE

Units;

h	-	hour
m	-	metre
s	-	second
kW	-	kilowatt
MW	-	Megawatt

Other;

BS	-	British Standard
CMI	-	Christian Michelsen Institute, Norway
ESD	-	Emergency shut down
NFPA	-	National Fire Protection Association, USA
LPG	-	Liquified petroleum gases
RHO	-	Radiant heat output (from a fire)

1 INTRODUCTION

1.1 Fundamental questions.
Fire and gas detection systems are a crucial part of safety systems in off-shore, and many on-shore, sites - or are they? They are required by law, and common sense tells us that they need to be effective, yet when incidents occur they rarely reduce losses. They can cause unnecessary shutdowns and are expensive to maintain.

This paper explores the problem and finds some disturbing answers. It is concluded that a fundamental review of the methods of specification and design are required. A solution is described relating the performance of fire and gas detection systems to the requirements of operators, legislation and other safety systems.

1.2 Problems.
A gas release or fire is often detected by personnel before a detection system responds. It is said that this is because people quickly become aware of paraffinic gas or smoke, but in fact, most incidents are probably smaller than we require a system to detect. The problem is to define the size of release that a system must detect. Not all detection failures can be attributed to an incident's being below detection level; a review of past significant incidents, where the system might have been expected to respond and limit losses, revealed that a remarkable 60% escaped detection. The analysis of these failures uncovered fundamental weaknesses in specification and design methods and resulted in the recommendations in this paper.

The performance of most safety equipment is well supported by codes and practices. Their use in specifications leaves little doubt in the mind of the operator or the designer about what performance is required. No performance codes exist for offshore fire and gas detection systems, so specifications should be (but rarely are) developed from first principles.

A typical specification will state that the system "..must detect the earliest presence of gas and incipient fire conditions anywhere .... and must comply with relevant legislation". This is very laudable but impossible to achieve. This ideal system cannot be built, nor can it be tested against such nebulous precepts. In effect the operator is saying that he does not know exactly what he wants and is attempting to saddle the designer with the responsibility of providing a system adequate for site safety and compliant with current legislation without being excessive in terms of cost.

No designer can accept this responsibility; legislative requirements do not form a specification, nor is it a designer's place to assess the adequate safety levels for any site. His design will be based on his opinion, offered in good faith but not matched to the operator's real needs.

The current approach to specification and design poses many problems:

- the designer's opinion rarely matches that of the Operator.

- the project manager has no firm grounds for rejecting the design and usually has to face delays and increased costs while a compromise is worked out.

- the intended performance is not recorded, so an operator has no basis on which to validate the acceptability of the system for audits including Formal Safety Assessments. He relies on expert (but necessarily subjective) opinion which might differ from the original design and thus incur further expenditure. Indeed, the Certifying Authority may have yet another opinion about what is acceptable. Another problem then is that there is no test or basis for assessing the adequacy of a system's performance.

1.3 Summary.

To summarise, weaknesses in many current systems and in existing specification and design methods include:-

- poor historical performance
- losses not minimised
- lack of performance specification
- project design cost and time-scale penalties
- acceptability based on opinion
- lack of appropriate tests
- design methodology not recorded for future reference.

1.4 Solutions.

It is clear that the operators must grasp the nettle and accept responsibility for specifying the required performance of a system. This is reinforced by Lord Cullen's recommendations on Formal Safety Assessment and Fire Risk Analysis for offshore sites. For onshore sites, Operators are required to develop HSE CIMAH cases. These studies may make assumptions about the role of F&G systems in limiting risks which will define performance goals for personnel safety. To these, the operator will usually need to add protection for the environment and for plant.

The method of performance specification introduced here was developed to cover all three aspects of risk - life, environment and plant. It incorporates a test for the acceptability of a design or an installed system, and creates a record of the performance requirement throughout the life of an installation.

Fire detection and gas detection, although related, have different objectives, and are dealt with separately.

2 FIRE

2.1 Significance

A significant fire is defined as one which could damage (either directly or through escalation) personnel, environment and plant, or cause unacceptable loss of production.

The specification must therefore ensure that the probability of detecting a fire is greater than the probability of suffering significant loss or damage. This must include damage incurred during the delay between detection and extinction or control. This delay is called the Effective Response Time (**ERT**).

The **ERT** is important in deciding when automatic action is required and what it must achieve. If significant loss is likely within the **ERT**, then immediate response must be invoked to prevent loss.

We can now define our aims as

- Any significant fire should be detected and an appropriate alarm raised.

- Any fire capable of causing significant damage within the **ERT** should trigger effective control actions.

2.2 Limits.
To establish the operational limits of a system we must examine how fires affect us, and the design of protection systems.

Fires of the order of many MW h must be prevented by plant design (isolation, inventory reduction etc.) if fire protection systems would be unable to prevent significant loss.

The effects of lesser, but still large, fires can be lessened by active and passive protection systems, but active systems may have to be initiated, by the fire detection system, which must also give an alarm.

Still smaller, are fires which pose no immediate threat of large loss, but if not checked could escalate or inflict direct damage. Provided they are alerted, personnel will be able to prevent significant loss or damage, either directly or by initiating fire fighting systems.

Smaller again, we reach a threshold where the fire is so unlikely to cause significant loss or damage that its detection can be left to personnel, provided that the site is manned. A fire may of course grow from one category to another, so any system must respond to escalation.

These categories hold for every area of an installation. Differing levels of fire size and duration, and the risks to life, environment and plant, will define the significance of an incident in an area.

2.3 Hydrocarbon areas.
The requirements for fire detection in hydrocarbon fire risk areas are not covered by any performance related code or standard. Any specification for system performance must be developed from basic principles.

Fires must be specified in a way that is explicit and helpful to designers. The technique of defining fire size by flame (or pan) area does not indicate its damage potential: a small cutting flame can cause

more damage than a large diffusion flame. The principal damaging
feature of a fire is the heat generated and Radiant Heat Output (**RHO**)
is not only appropriate, it is easy to measure during tests of fires
and detectors.

It is necessary to define what maximum **RHO** of fire can exist in
every area without being detected, taking into account the fire's
duration and escalation potential, the nature of the release, risks to
life and vulnerability of plant, among other things. Obviously,
detection of smaller fires is desirable, this is an upper limit.

This **RHO** limit will vary in each area, and unfortunately incident
records give meagre information on the levels of **RHO** that have led to
damage, although no significant hazard appears to exist in fires
smaller than 10 kW RHO (this is about 0.2 m² of diffusion flame).

In hydrocarbon plant we can define three levels of required system
performance (Grades):

Grade A - In the highest risk areas, fires of 10 kW RHO or larger
could lead to damage or escalation within the **ERT**. They therefore
need the system to raise an alarm and also initiate effective
action.

Grade B - Most areas of hydrocarbon plant can withstand a 10 kW RHO
fire for longer than the **ERT** before significant damage is likely, so
the system should raise an alarm but does not need to initiate
automatic controls unless the fire reaches say 50 kW RHO (or starts
at or above this **RHO**).

Grade C - Some areas contain no stored inventory and, even if a fire
occurred there, nothing important would be damaged. Alarms are
required only when the fire reaches about 100 kW RHO; action is
needed when the fire is large enough to threaten Grade A or B areas,
say 250 kW RHO.

Each item and part of the hydrocarbon plant should be reviewed and
categorized as Grade A, B or C. This tells the designer that no fire
of the specified **RHO** or larger should occur within the Graded area
without being detected, all parts of that area must be sufficiently
close to detectors to achieve the required response. (Note, it is
quite permissible for a smaller fire to initiate the same responses.)

2.4 Specification for Hydrocarbon areas.

A specification for fire detection in hydrocarbon areas must define
several parameters, and the first should relate the Grades to **RHO**
thresholds for alarm and action. These may vary from site to site, but
for offshore production installation the values given in Table 1 are
generally appropriate.

TABLE 1   Threshold Fire **RHO** by area Grade

	Fire Output (kW RHO)	
	Alarm	Action
Grade A	10	10
Grade B	10	50
Grade C	100	250

The maximum time to alarms/action after ignition must be specified; a value of 10s is practical without being a significant part of the ERT.

A value must be set for probability of response. The greatest single influence in achieving this value is whether a fire will be exposed to detectors or screened by obstacles. A value of around 0.85 is suggested, since 10 kW fires are believed less likely to cause damage. Obviously, if a fire grows, so too does the probability of damage, and a higher probability of detection becomes necessary. Common mode factors which affect response to large and small fires alike, e.g. the control system, must have a high probability of correct reaction.

The specification must define where Grades are to be applied, by listing all items of hydrocarbon plant and assigning Grades to each.

Grade A - should have a 1 metre surround to encompass thermally weak items, e.g. small bore pipe work associated with the risk item.

Grade B - should have a 2 metre radius to detect fires encroaching from Grade C areas. Grade B areas always surround Grade A areas, and may also be used for specific items of plant.

Grade C - All parts of a hydrocarbon area must be Graded. Any which are not Grade A or B must be Grade C. In enclosed areas Grade C has no radius but extends to the walls; in open areas a 2 metre radius is thought sufficient.

Plan drawings showing Graded plant are useful for illustrating the listing and, as discussed later, for checking design adequacy. They are however merely analogies of what is essentially a three-dimensional problem and must therefore be treated with judgement.

To ensure that appropriate detectors are used, the specification must include the types of fire which might arise, and the fuels that might be present.

This specification forms a permanent record for future reference. The detailed information is a suitable basis for assessments and the design of later modifications.

2.5 Effective range of flame detectors.

Flame detectors are commonly used in hydrocarbon areas, but they are not governed by standard tests directly related to offshore applications. Each manufacturer performs his own unique tests and the quoted ranges for detectors do not take into account the types of fire, environment, or the condition of the detector on the site - all of which can seriously reduce a detector's range. The 'Effective Range' of flame detectors in service on hydrocarbon production sites may be half of the quoted range, thus the designer may seriously over estimate the ability of the system.

An economical test has been developed to estimate the Effective Range. The resulting values relate Effective Range directly to the RHO of hydrocarbon fires and offshore conditions, so ensuring that a detector in service is as sensitive as assumed in the design, given adequate maintenance. These Effective Ranges replace the more optimistic values of the manufacturer, and must be included in the specification.

2.6 Non-hydrocarbon areas.

Codes and Standards exist for fire detection in general onshore applications which define appropriate detectors and installation procedures. It is tempting to use them for non-hydrocarbon areas of offshore sites, e.g. offices and accommodation but they have limitations because they assume different loss risks.

Most current codes (e.g. BS 5839 part 1, 1988 and NFPA 72E 1990) are derived from tests by Factory Mutual in the USA which laid the foundations for today's onshore fire detection. A major conclusion of the original work, still relevant today is that fires of less than 250 kW RHO are difficult to detect using temperature or smoke detectors. This is because the relatively small quantities of heat and smoke produced are easily affected by air movement and thermal layers. The products of larger fires are more likely to overcome these masking effects and reach the ceiling mounted detectors. The most commonly applied code (BS 5839 Part 1, 1988) draws attention to these problems but it offers no solutions. Nor does any other code, so either 'opinion engineering' is applied or, perhaps worse, codes are used where they are not valid. The subject of the size of fire to be detected is not addressed.

These limitations do not generally affect onshore applications, but would be serious indeed if this type of fire (about the size of a door) arose in a sleeping cabin or ESD equipment room. Specifications must give clear guidance for areas where Codes might not apply.

Non-hydrocarbon areas can also be Graded into three categories by examining the likely consequences of a fire, as follows.

Grade D (Life) - applied to areas where personnel sleep or are at particular risk (e.g. sickbay), including associated escape routes and ventilation systems. We recommend the use of BS 5839 for type L1 requirements, plus installation of sprinklers to the NFPA code (NFPA 13,1987). According to the BS Code, type L1 will protect lives in areas adjacent to, but not necessarily within, rooms. This will depend on detector siting relative to the fire and draughts from the ventilation system. The sprinklers do not of course aid detection, they are unlikely to respond before smoke detectors, but their installation will help to control and contain the fire.

Grade E (Equipment) - applied to any area containing safety systems or emergency facilities and equipment, including system panels, power supplies and, where necessary, related cable routes. Here the aim is to detect fires before the system can be damaged. The best available code is BS5839 (type E1 systems), not BS6266 Appendix 1 as often supposed. The current revision of BS6266 might make it more helpful.

Equipment areas are frequently well ventilated, or hot, or both. This makes early fire detection difficult and non-standard methods have to be specified, as discussed in the next section.

Grade F (General) - applies to all other non-hydrocarbon areas where a damaging fire is possible, i.e. areas where loss of contents is acceptable so long as ready control is ensured (offices, flammable material stores, workshops, recreation areas etc.). The aim is to detect fires before they can spread or become difficult to control. The BS5839 requirements for type E2 areas are normally adequate for these areas.

2.7 Small fires in Non-Hydrocarbon areas.

It is likely in Grades D and E areas that fires smaller than 250kW should be detected. Fires which can damage personnel and critical equipment may be small, with weak plumes. This problem cannot be fully addressed in the space allowed in this paper but a brief discussion follows.

Prediction of air flow patterns is unreliable and no better method has been found for protection of sleeping cabins etc. than to trace air flow with cosmetic smoke, then site detectors accordingly. Duct sampling with sensitive detectors may also be a solution but is likely to raise unwanted alarms.

In larger spaces such as switchgear rooms, beam-type smoke detectors have proved the most effective, and have been arranged to monitor air flows which entrain the smoke. Extra point detectors might be needed in areas of poor ventilation, pinpointed by trace testing. Alternative approaches are the use of detectors in cabinets, duct sampling and Very Early Smoke Detection Apparatus (VESDA) systems.

The effect of heat barriers is always underestimated. Smoke plumes from small fires cool quickly in the resulting turbulence, mixing to a density similar to or heavier than air, and may not reach ceiling mounted detectors if a thermal layer of just a few degrees above ambient temperature exists. Again, cabinet, duct and VESDA systems could be considered.

In summary, any or all of the following might need to be employed:
Air movement surveys
Beam detectors to monitor entraining air flows
Point detectors in low air movement areas
Duct mounted detectors (rarely used)
VESDA systems (often used for critical equipment).

2.8 Specification for Non-Hydrocarbon areas.

To ensure an optimum design, the following information must be included in the specification:
- Grade definitions, using appropriate Codes
- Code limitations
- List of areas requiring detectors, and including
Grade of area
Ventilation rates
Nature of risk(s)
Classes of fire requiring detection
- Provisions for areas which exceed Code limitations. (normally requiring an air movement survey before final siting and commissioning of detectors.)

During commissioning, extra beam, duct, point and VESDA systems should be available to cope with the unexpected, as should spare cabling to a control panel with available capacity.

2.9 Assessment.

The responsibilities of the designer can now be clearly defined. He will be expected to
- divide the site into zones
- select appropriate detectors
- plan their numbers and distribution
- structure the system to give the alarm and action outputs
- define input requirements for control panels
- define installation and commissioning requirements
- recommend maintenance schedules.

(He may also be asked to review for other possible hazards, but this should not be necessary.)

The production of the optimum design no longer depends on subjective opinion, and its assessment has become a test of design against specification.

Design intent is often misunderstood or overlooked during commissioning. Problems occur in the module yard or hook-up when detectors are installed incorrectly, or when other items such as ventilation grilles, cable trays and pipe-racks are not installed exactly as the designer envisaged, therefore the Operator must employ an effective checklist which includes:
Smoke and Heat detectors;
- ensure coverage meets the specifications
- visual inspection of detectors
- air movement testing
- performance testing of detectors
Flame detectors;
- performance testing of detectors
- photographic survey of flame detector coverage
- effectiveness of control actions

A useful tool has been developed for flame detector coverage assessment. When planning the layout of flame detectors, it is probably assumed that each will have a clear line of sight to the limits of its Effective Range (or to the walls) but this seldom happens; a flame might be obscured by plant and fixtures. A photograph should be taken from each detector location, and the footprint of its actual view edited in a computer model. A program is run which compares the requirements of the Grading with the footprints of the detectors. From this, potentially weak areas can be identified.)

The audit procedures described above afford the best available assurance that a system will be capable of protection at the specified levels. There is no absolute test of a fire detection system's capability except to create fires, which may only possible during onshore construction, not when plant is operational. Despite much R&D effort, no way has yet been found to simulate a calibrated fire in an operational area.

Whenever there are changes to the risks or consequences of fire in an area, its Grading must be reviewed, because a change in Grade might necessitate modification of the detection system.

3 GAS

3.1 Significance.
Losses may arise from many types of release, but this discussion covers only combustible leaks. It should be remembered that other forms of release detection may be required.

A significant release of gas is defined as one which may feasibly cause
- death or injury of personnel
- environmental damage
- unacceptable loss of production or plant.

Combustible gases are not usually damaging until ignited, when the result might range from a flash fire to explosion. Most fires will be controlled by the fire detection system, but explosions may cause damage too quickly for conventional detection to help. Explosion damage occurs from

- over-pressure: the pressure developed between the expanding gas and its surrounding atmosphere.

- pulse: the differential pressure across plant as a pressure wave passes might cause collapse or movement.

- missiles: items thrown by the blast of expanding gases might cause damage or escalation.

A good introduction to the subject of vapour cloud explosions is given in Harris 1989.

In plant areas, these effects can be related to flame velocity, but where this is below a threshold of about 100 m/s, damage is unlikely (within the limits of confinement normally found offshore). This leads to the question of the size of cloud or plume in which such velocities can occur. Early CMI work (Hjertager et al 1988) demonstrated that flames need a 'run-up' distance of around 5.5m to reach damaging speeds thus clouds with smaller dimensions are not likely to cause damage. This is a drastic oversimplification of the variables involved, but assumes the worst likely cases of congestion, confinement and gas concentrations.

If, as CMI work suggests, 5.5m is the threshold for damage, this would support the siting of detectors in a three dimensional triangular grid of 5.5m. Designs should therefore be planned on a 5m grid, to give some freedom of movement for commissioning adjustments. This would involve numerous point detectors and re-poses the question, Can numbers be reduced? Reasons why they should not are

- Increasing the 5.5m value to 6 or 7 does not significantly reduce the number of detectors (this is a function of the typical sizes of modules), but does significantly increase the volume of cloud and size of plume that might remain undetected.

module means that a flammable concentration exists and detector siting will affect the speed of response more than calibration or alarm levels.

To achieve early and reliable warnings of leaks, the sensitivity of detectors should be set at the highest level commensurate with acceptable false alarm rates. For pellistor devices, the life of the detector is taken into account, so the increase in gain (i.e. calibration factor) is just above the actual calibration for the detection of Propane; with more stable instruments, higher gains should be possible.

Beam detectors are criticised because they cannot indicate gas concentrations at any point, only the amount of gas present in the beam. In reality, they meet operational requirements better than point detectors, and their sensitivity and reliability are claimed to be better than pellistor types.

3.3 Control actions.

In the report on the Piper Alpha disaster (section 19.41), Lord Cullen observed that gas detection systems may not be making their full contribution to protecting against leaks which may cause serious explosions, particularly in initiating ESD actions. In practice, automatic actions triggered by gas detection are rarely effective in reducing the risk of damage. A principal reason is that plant block and depressurising actions do not stem the release quickly, the release may continue beyond the 'blow-down' time of the plant. The cloud has ample time to grow and engulf an ignition source, so there is an argument here for selective plant de-pressure but this is rarely done.

A useful measure, often disregarded, is to reduce ignition sources. Some sites appear to have better ignition to release ratios than others. This implies that control of all electrical equipment, and automatic isolation of utility outlets (electrical, welding, and especially pneumatic) is effective.

Water systems offer hope. Recent British Gas work on the mitigation of explosion by water sprays (Acton R. 1990) brings a prospect of directly reducing flame speed, a principal damage factor. Work is still needed to optimise the effectiveness of these special spray systems, and ensure that they are not themselves an ignition source. Conventional deluge systems are not reliable for dealing with gas clouds; evidence suggests they can increase turbulence (and therefore damage) and cause ignition. There is no evidence for mitigation except the reduction of damage from consequential fires.

Explosion suppression systems are being offered based on powder or Halon extinguishants but they have a big weakness. We have already shown that releases may continue for some time and an ignition source, if present is not likely to disappear. Re-ignition of the gas cloud is likely with such 'one-shot' systems.

Gas detection is often installed in ducts with the aim of preventing a flammable mixture reaching the protected space (or engine). In practice, designs are fundamentally flawed because the

- The 5.5m distance is based on hard evidence and difficult to fault. It can be argued that levels of congestion and confinement in many areas are below those of the tests, but other effects, notably ignition energy, may be severely heightened. The variables involved are not well understood and more research is required before data are reliable. Most work on part-filled areas (e.g. Lord Cullen' report on Piper Alpha 6.23 to 6.43, and Buckland 1980) does not justify optimism.

- Fewer, but highly sensitive detectors could be substituted on the assumption that the concentration gradients at the edge of plumes and clouds are gradual. But this assumption is wrong: the boundary between air and average gas concentration can be very sharp.

- It has been suggested that the number of detectors might be reduced if their layout was specific to the plant layout and ventilation patterns in an area. During the development of the suggested specification, several installations were critically reviewed using cosmetic smoke trials to determine optimum detector location and distribution. Results showed that nothing could be gained by 'custom design' of detector distribution in naturally ventilated areas: too many variables exist, including location of leaks, strength and direction of air currents, and the effect of plant layout on a plume.

- As discussed later, automatic control actions will take time to reduce the probability of a serious fire or explosion. The earliest possible but reliable action is required. This is a function of detector spacing.

There are reliable ways to reduce the implications of these findings. Linear beam instruments might reduce the numbers of point detectors required. They could also be used to monitor the boundary of an area containing sources of potential release and eliminate the need for detectors in an adjacent source-free area.

In open areas (e.g. top-deck plants or on-shore sites), high over-pressures are unlikely unless a large cloud engulfs major plant items which might act as bluff body turbulence generators and accelerate flame velocity - as happened at Flixborough. There is no direct evidence from tests or accidents that shows the minimum conditions likely to cause such an event, but the indications are that several large items (or equivalent congestion) would have to be involved before high flame velocities were likely. In severely congested spaces within open plant, damaging over-pressures could result from smaller releases, as indicated in experiments by Shell at Spadeadam. Unfortunately, these results have not been published.

Adequate protection may be provided for open areas by a beam detector across the top of plant, or perimeter detection for liquid phase releases of condensates etc., both of these being supplemented with point detectors in heavily congested spaces.

3.2 Calibration.
It is widely believed that operators need to know gas concentrations at different points in plant areas; in fact, what they need is immediate knowledge of a significant leak. Neat gas released anywhere in a

reaction time of the detector and control action greatly exceeds the transit time of the gas in the duct. Special detectors (or excessively big ducts) are required if such designs are to work.

3.4 Specification.

A specification for gas detection system performance should define the size and nature of likely hazards in each area. In confined or enclosed areas, the system should raise alarms to a cloud or plume of 5.5m diameter or more. Immediate action should be triggered to reduce the probability of ignition and locally de-pressurise the source of release.

In open areas, the system should alarm the presence of an engulfing cloud, with extra detectors in congested or confined areas.

Detection of gas in ducts serving safe areas and engines should be designed to prevent gas from entering the protected space.

3.5 Assessment.

The responsibilities of the designer can now be defined as;
- select appropriate detectors
- plan their numbers and distribution
- structure the system to give the alarm and action outputs
- define input requirements for control panels
- define installation and commissioning requirements
- recommend maintenance requirements.

The assessment of a gas detector layout should ensure that
- The spacing between detectors meets the specified minimum
- All detectors are accessible for maintenance
- Beam types are arranged so that they are not likely to be interrupted
- Control actions and alarms meet the intent of the specification.

In a similar way to the assessment of fire detection adequacy, check procedures are required at design review, commissioning and during the life of the site.

4 CONCLUSIONS

4.1 General.
- Operators and designers have failed to develop fire and gas detection systems based on perceived hazards and risks.
- The performance of fire and gas detection systems is usually lacking in their ability to protect against loss.
- Most weaknesses can be attributed to poor definition of requirements and lack of understanding at the design stage.
- Loss prevention by these systems can be significantly improved by employing existing knowledge and the findings of this paper.

4.2 Fire detection.

- All areas should be 'Graded' with respect to the risks and consequences of fires.
- The Grades should define the required performance of the detection system.
- The 'Effective Range' of flame detectors must be established for the intended risks and conditions of operation.
- The current Codes used for fire detection in non-hydrocarbon risk areas do not offer adequate protection for many offshore applications. Special attention is required to the design of systems in these areas.

4.3 Gas detection.

- Gas plumes or clouds of 5.5m diameter are a significant hazard in some areas. Detectors should be arranged to detect them.
- Gas clouds which engulf significant plant volumes in open areas should also be detected.
- Control actions to gas releases could be made much more effective in preventing loss with current technology. Even better methods are being investigated.
- Calibration of gas detectors to specific hazards in general areas may reduce system effectiveness.
- Beam detectors have many applications offshore.

REFERENCES

Acton, R., Sutton, P. and Wikens, M.J., (1990), 'An investigation of the mitigation of gas cloud explosions by water sprays', *I. Chem. E. Symposium Series No. 122*, pp 61-76.

British Standards Institution, (1988), BS 5839:Part 1,, *Fire detection and alarm systems for buildings: Code of practice for system design, installation and servicing.*

Buckland, I.G. (1980) 'Explosions of gas layers in a room sized chamber', *I. Chem. E. Symposium Series No. 58*, pp 289-304.

The Hon Lord Cullen, (1990), *'The public inquiry into the Piper Alpha disaster'*, HMSO, London

Harris, R.J. and Wickens, M.J. (1989) 'Understanding vapour cloud explosions - an experimental study', *I. Gas Eng. communication 1408.*

Hjertager, B.H., Fuhre, K. and Bjorkhaug, M. (1988), 'Concentration effects on flame acceleration by obstacles in large scale Methane - air and Propane - air vented explosions', *Combust. Sci. and Tech.*, Vol. 26, pp. 239-256.

National Fire Protection Association, (1987), NFPA 13, *'Standard for the installation of sprinkler systems'.*

National Fire Protection Association, (1990), NFPA 72E, *'Standard on automatic fire detectors'.*

SESSION D:PROTECTIVE SYSTEMS

An Overview of draft IEC International Standard:

"Functional Safety of Programmable Electronic Systems"

R Bell
Health and Safety Executive
Technology Division
Magdalen House
Bootle, Merseyside L20 3QZ
UK

S Smith
B H-F (Triplex) Inc.
20316 Gramercy Place
Torrance
California 90501
USA

1. INTRODUCTION

1.1 General

Computer-based systems, generically referred to as Programmable Electronic Systems (PESs), are increasingly being used in safety-related applications. This trend is certain to continue because of the potential advantages such systems offer. The potential safety advantages will, however, only be realised if appropriate design and assessment methodologies are used. Unfortunately, many of the features of PESs do not enable the safety integrity to be predicted with the same degree of confidence that has traditionally been available for less complex hardware-based systems.

Several bodies have either published, or are developing, guidelines to enable the safe exploitation of this technology. In the UK, the Health and Safety Executive (HSE), a major safety regulatory authority, developed guidelines for programmable electronic systems used for safety-related applications. The guidelines were developed taking into account work going on within other countries and after considerable research and discussions with industrial interests. The guidelines adopt a systems approach and provide a methodology that enables a systematic approach to be adopted to both the design and assessment of systems incorporating programmable electronics.

The guidelines were published in June 1987 under the general title "Programmable electronic systems in safety-related applications". Two documents were published[1]:-

1) "An introductory guide".

2) "General technical guidelines".

The guidelines are generically-based and should enable the safety integrity of systems incorporating PESs to be determined irrespective of the application.

In the Federal Republic of Germany a preliminary draft standard DIN V VDE 0801[2,3] has recently been published. Also, within the European Community an important element in the work on harmonised European Standards, in connection with the requirements of the Machinery Directive, is concerned with safety-related control systems (including those employing PESs.[4].)

In the USA, the Instrument Society of America (ISA) is well advanced in preparing a standard on "Programmable Electronic Systems for Use in Safety Applications". The Centre for Chemical Process Safety (CCPS), a Directorate of the American Institute of Chemical Engineers, is also preparing guidelines for the chemical process sector.[5]

It is important to ensure that guidelines development takes place taking account of the work going on within the standards organisations at both national and international levels and that guidelines that are developed nationally are progressed internationally. A major objective is the achievement of international standardisation. In this context the International Electro-technical Commission (IEC) is developing two International Standards and has issued two draft International Standards titled:-

1) Functional Safety of Programmable Electronic Systems: Generic Aspects; Part 1: General Requirements. This draft standard ("systems" draft standard) was prepared by Working Group 10 (IEC/SC65A/WG10).[6]

2) Software for Computers in the Application of Industrial Safety-Related Systems. This draft standard was developed by Working Group 9 (IEC/SC65A/WG9).[7]

Both of these draft International Standards were sent to all National Standards Committees for them to comment upon. Consultation was seen as a necessary and important step in the production of robust International Standards in these two important areas. The consultation period finished in early 1990.

1.2 Systems draft standard

The Scope of the systems draft standard[6] is wide ranging and applicable to all application sectors and, when finalised, could form the foundation upon which other application standards could be based. This should lead to a high level of compatibility between standards covering different application sectors. Such compatibility has important safety and economic advantages.

The draft standard provides guidelines on the aspects that need to be addressed when programmable electronic systems (PESs) are used to carry out safety functions. A PES is defined as:-

"A system based on one or more central processing units (CPUs), connected to sensors and/or actuators, for the purpose of control, protection or monitoring" ; (seeFigure 1).

The draft international standard was developed:-

- To enable the functional safety of PESs used in safety-related applications to be sys tematically addressed.

- With a rapidly developing technology in mind.

- To allow future development of the Standard to reflect changes in safety integrity criteria and assessment techniques.

- To allow future application - specificInternational Standards to be developed.

This Chapter gives an overview of the systems draft standard and :-

- **explains its intended purpose ;**

- **highlights the key underlying features ;**

- **considers the way-ahead.**

2. GENERIC AND APPLICATION SPECIFIC STANDARDS

In the past the various application sectors (eg manufacturing, medical, process, transportation etc) have developed their own standards for safety-related systems. These have evolved in a specific, and sometimes unique, manner with little in common with other application-sector standards. With the advent of programmable electronic devices and their incorporation into systems (programmable electronic systems (PESs)) a significant problem arose in all sectors concerning guidelines for safety-related systems. For example, there would be serious resource problems if each application sector were to develop its own unique approach in developing standards for its application sector. It is therefore better to develop a common, or generic, approach that can be used across all application-sectors. The concept of a generic international standard leading to the development of application-specific international standards is not new in the standards field. This concept is shown in Figure 2.

The draft standard is generically based and a major objective is that it be used by other application sectors to develop application-specific standards. Adopting this approach enables the application-specific standards to be based on the same underlying principles. This will have many benefits to those who may be involved with more than one application sector eg. equipment suppliers. It also means that future work done at the generic level will benefit all application sectors.

Figure 1 Diagram to Illustrate PES Structure and Notation

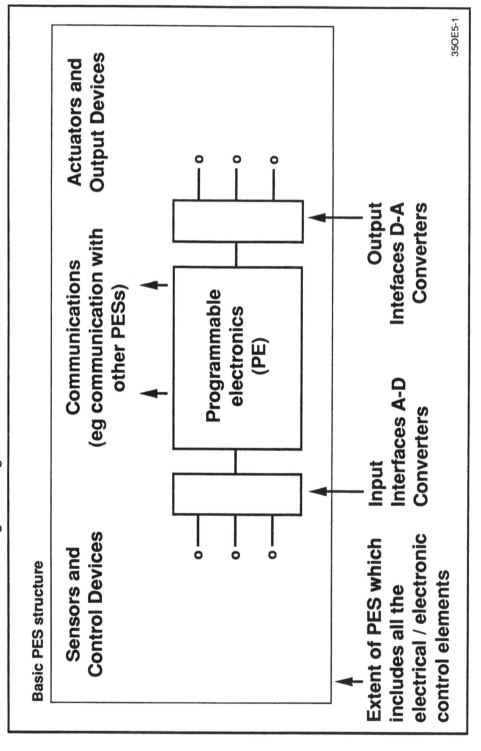

Basic PES structure

35OE5-1

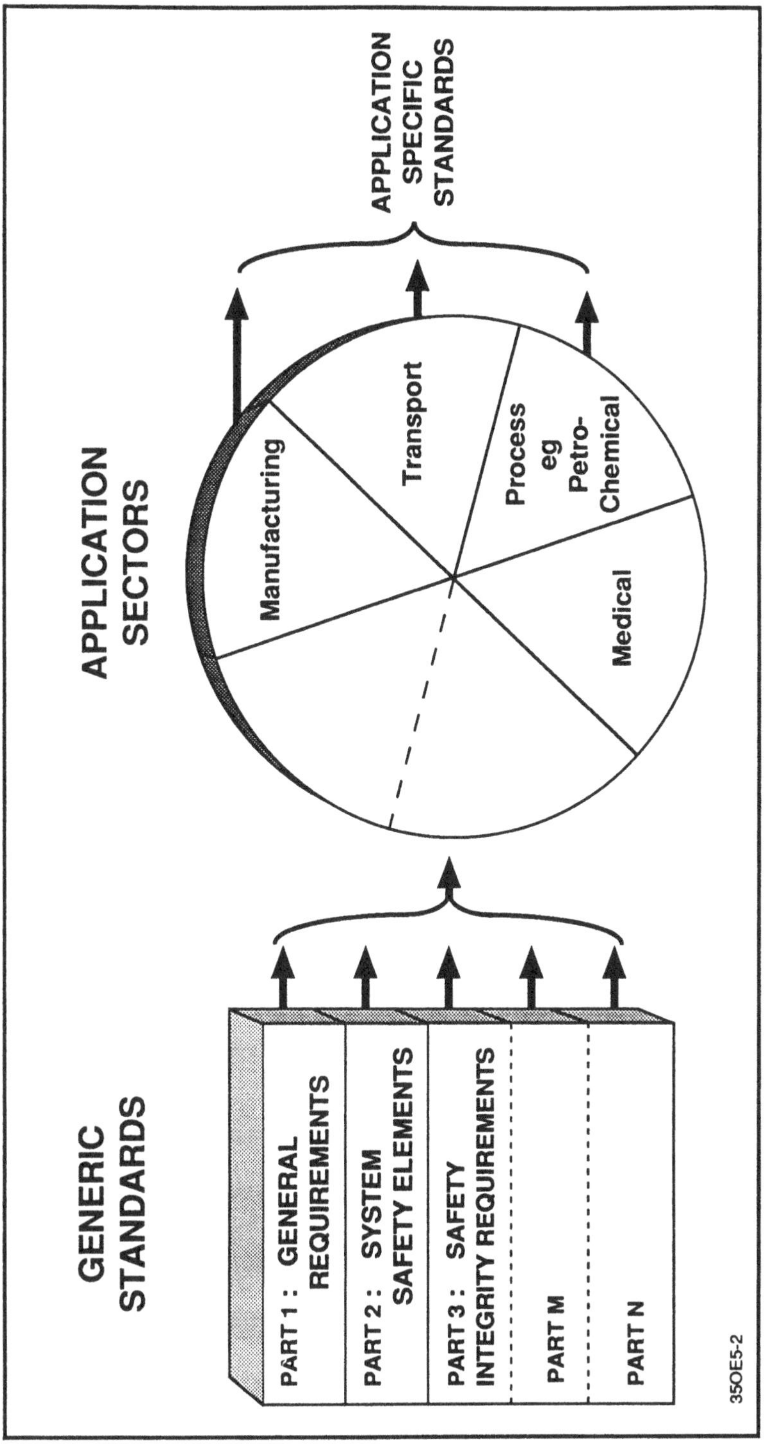

Figure 2 Generic and Application - Specific Standards

GENERIC
STANDARDS

APPLICATION
SECTORS

APPLICATION
SPECIFIC
STANDARDS

PART 1 : GENERAL REQUIREMENTS

PART 2 : SYSTEM SAFETY ELEMENTS

PART 3 : SAFETY INTEGRITY REQUIREMENTS

PART M

PART N

Manufacturing

Transport

Process
eg
Petro-
Chemical

Medical

35OE5-2

The draft standard is Part 1 of a series and a key feature is that each part will deal, in a generic manner, with specific aspects. It is proposed that future Parts will cover:-

- hazard analysis
- risk assessment
- system safety elements
- safety integrity requirements
- system integrity levels
- system validation
- retro-fitting
- documentation

3. SCOPE

The draft standard is primarily concerned with safety-related systems incorporating programmable electronics or other complex electronic devices and lays down a framework that could be applied to safety-related systems irrespective of the technology on which those systems were based eg (electro-mechanical, solid state electronic, programmable electronic, pneumatic or hydraulic etc).

Note: National Standards Committees were requested, during the consultative phase, to indicate whether they would prefer the draft standard to concentrate on PESs or cover all types of safety-related systems (irrespective of the technology on which they were based). It was agreed that although the proposed International Standard should concentrate in detail on electrical/electronic/programmable electronic systems, it should also contain the general methodology applicable to all types of safety-related systems. The overview given in this Chapter concentrates on PESs since this was the main concern in the draft standard.

The draft standard :

- applies to safety-related systems when one or more of such systems incorporate pro grammable electronics or other complex electronic devices.

- is generically based and can be used for both design and assessment purposes irrespec tive of the application. Examples of the application sectors coming within the scope of the Standard include:

- process industries (emergency shut down systems, fire and gas detection systems, boiler controls).

- manufacturing industries (industrial robots, machine tools).

- transportation (railway signalling, braking systems, elevators/lifts).

- medical (electro-medical equipment eg. (radiography)).

- is mainly concerned with safety to persons but is also applicable to environmental issues.

- applies to the total configuration of safety-related systems (ie the total hierarchy of safety-related systems).

- uses a Safety Lifecycle Model for all activities necessary for ensuring that the required safety-integrity levels are met for the safety- related systems under consideration.

- addresses all relevant aspects associated with the design and assessment of safety-related systems.

- has been developed to enable the responsible Technical Committees in Standards organisations to develop application-specific International Standards.

- provides guidance on the application of PESs having safety functions in fields where no specific recommendations are available.

- introduces the concept of "safety integrity levels" for specifying the level of safety integrity of safety-related systems.

Note: It is not the intention in the proposed International Standard to provide guidelines for classifying the safety integrity level required for a specific application. This must be established based on knowledge of the application sector or other related experience. It will be for the responsible Technical Standards Committees, dealing with specific application sectors, to specify the safety integrity levels for associated tolerable risk levels.

4. SAFETY INTEGRITY AND RELIABILITY TERMS

The difference between the terms **Safety Integrity** and **Reliability** have special meanings within the context of the guidelines. Consider the circle, in Figure 3, comprising two sectors, A and B. The total area of the circle represents failures from all causes.

- Area **A** represents those failures due to **Random Hardware Failures**. That is, hardware failures resulting from various breakdown mechanisms that occur at unpredictable (ie, random) times.

- Area **B** represents **Systematic Failures**. That is, failures that are due to errors that have been made at some stage in the specification, design, construction, operation or maintenance of a system. Systematic Failures are particularly important in the context of complex systems and are therefore particularly relevant to PESs.

Figure 3 Safety Integrity and Reliability

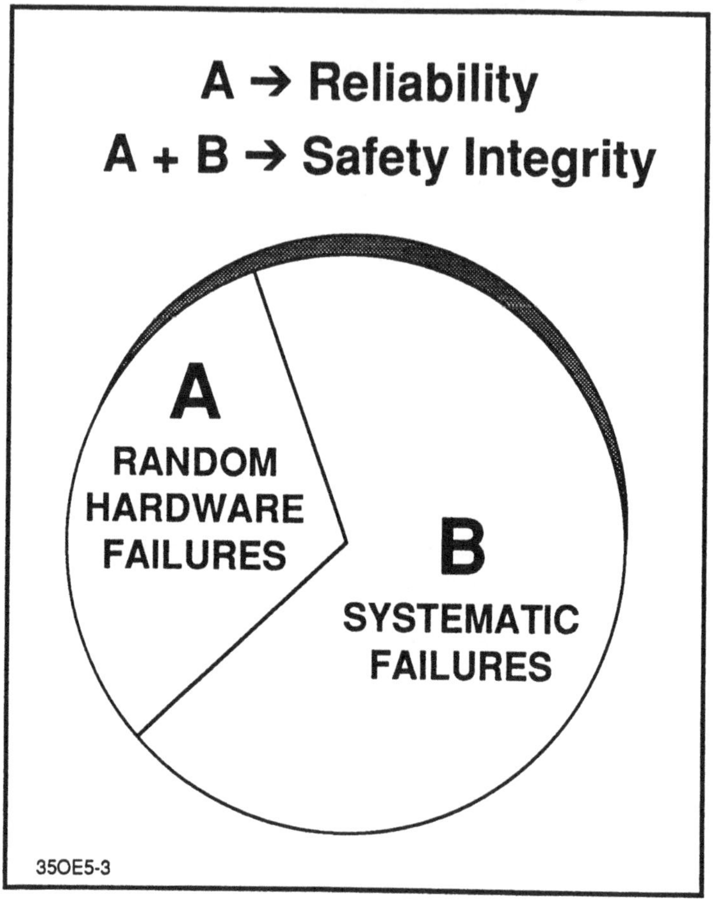

The term **Reliability** is used to indicate the degree of precautions built into a system against Random Hardware Failures. In particular, those failures that lead to a potentially dangerous situation. The term, as used for the purpose of the draft standard, is narrower in meaning than is commonly used relating only to hardware failures in a dangerous mode of failure. In common usage the term includes failures in both safe and dangerous modes.

The term **Safety Integrity** is used to indicate the degree of precautions taken against all failure causes (both Random Hardware Failures and Systematic Failures). Safety-Integrity is therefore affected by such things as errors or omissions in the safety requirements specification, the integrity of the software, the immunity of the safety-related systems to electrical interference and the way in which the system are maintained.

Random Hardware Failures can be quantified to an acceptable degree of accuracy using reliability prediction techniques. Unfortunately, **Systematic Failures** cannot be predicted in a quantified manner to any acceptable degree of accuracy and qualitative measures have to be adopted.

5. GENERIC SAFETY PRINCIPLES :

To ensure safe operation of safety-related PESs, it is necessary to recognise all various possible causes of PES failure and to ensure that adequate precautions are taken against each. The draft standard categorises the types of failure as follows:

- **Random hardware failures; and**

- **Systematic failures.**

Any strategy for safety-related PESs must consider both categories of failure. The strategy adopted in the draft standard for dealing with dangerous mode failures, of the safety-related systems, is based on three **system safety elements** (system characteristics) which are intended to provide adequate precautions against both random hardware and systematic failures. The system safety elements are:
- **Configuration;**
- **Safety-related hardware reliability and**
- **Systematic integrity**

The exact "package" of system safety elements will depend on the level of safety integrity to be achieved and therefore on the specific application.

Note: It is proposed to develop minimum Technical requirements for the system safety elements for each system Integrity Level in future parts of the proposed International Standard.

6. SAFETY-RELATED SYSTEMS

The concept of a safety-related system is fundamental to the draft standard. It is defined as a system that:-

- implements, independently of any other system, the required safety functions necessary to achieve a safe state for the Equipment Under Control (EUC) or to maintain a safe state for the EUC.

- achieves, on its own or with other safety-related systems, the necessary level of safety integrity for the implementation of the required safety functions.

The term safety-related system refers to those systems that enable, independently of other systems, the tolerable risk level to be met. Safety-related systems can broadly be divided into two types; control systems and protection systems. It should be noted that not all control systems will necessarily be designated as safety-related systems. The required level of safety integrity may be achieved by implementing the safety functions in the main control system of the EUC or the safety functions may be implemented by separate and independent systems dedicated to safety (eg protection systems). It should be noted that a person could be part of a safety-related system.

Those systems that make up the hierarchy of systems for control, protection or monitoring are termed the "total configuration of systems". All the safety- related systems which, together, achieve the necessary level of safety integrity, are termed the "total configuration of safety related systems".

The principles developed in the draft standard apply to the total configuration of safety-related systems required to achieve an adequate level of safety integrity to meet the tolerable risk and will comprise, in many cases, both PES and non-PES safety related systems.

7. RISK AND SAFETY INTEGRITY

The draft standard adopts a total systems approach and part of this involves the separation of the **risk of the EUC** from the **safety integrity of the safety related systems.** The terms risk and safety integrity are defined as follows:-

RISK: The combination of the frequency, or probability, and the consequence of a specified hazardous event. *(Note: The concept of risk always has 2 elements; the frequency, or probability, with which a hazard occurs and the consequences of the hazardous events.)*

SAFETY INTEGRITY: The likelihood of a safety-related system achieving its required functions under all the stated conditions within a stated period of time.

In determining the safety integrity level achieved, the precautions taken against all causes of failures (both systematic failures and random hardware failures) which lead to an unsafe state should be considered. Random hardware failures may be quantified using such measures as the failure rate in the dangerous mode of failure or the probability of the protection system failing to operate on demand. However, the safety integrity of a system also depends on many factors which cannot be accurately quantified and have to be considered qualitatively.

In any particular application there will be a risk level that can be deemed tolerable. For situations where the level of risk, without precautions, is not tolerable, there is a need to ensure that the applied safety-related systems reduce the risk to the tolerable risk level. This general concept is shown in Figure 4. It can be seen that the tolerable risk level is met, for example, by reducing the probability of the hazard occurring by adopting safety measures which include safety-related control systems (eg protection systems). For example, with respect to Figure 4, the tolerable risk level is achieved by reducing the risk level from A to D by the addition of safety-related control systems. The minimum reduction being A to C, but the actual reduction being A to D.

The purpose of the tolerable risk level is to state that which is deemed tolerable with respect to both the frequency (or probability) of the hazardous event and its consequences of it. It is the EUC that creates the hazard. The safety-related systems are designed to reduce the frequency (or probability) of the hazardous event and/or the consequences of the hazardous event. That is, the safety-related sytems are designed to achieve the required risk reduction (ΔR) to enable the tolerable risk level to be met (See Figure 5).

Safety integrity relates to the performance of the safety-related systems in carrying out the safety functions. The required safety integrity level, is specified in the Safety Integrity Requirements Specification and is formulated in terms of the three system safety elements. (See Figure 7)

Tolerable risk levels are usually set by interested parties within the application sector in question. That is, they are determined on a societal basis. Safety integrity is an engineering concept which allows the safety- related systems to be specified in engineering terms. The safety integrity level, for the total configuration of safety-related systems, needs to be appropriate to meet the tolerable risk level, and is set by the application sectors in question.The separation of risk and safety integrity is shown in Figure 5.

8. SYSTEM AND SOFTWARE INTEGRITY LEVELS

The draft standard adopted the concept of **system** Integrity Levels and **software** Integrity Levels. The draft standard on systems did not specify the number of Integrity Levels (for systems) but it is probable that 4 levels will be adopted. In order to aid understanding of the concepts, 4 Integrity Levels for the systems are assumed in the rest of the Chapter.

280

Figure 4   Risk Reduction: General Concepts

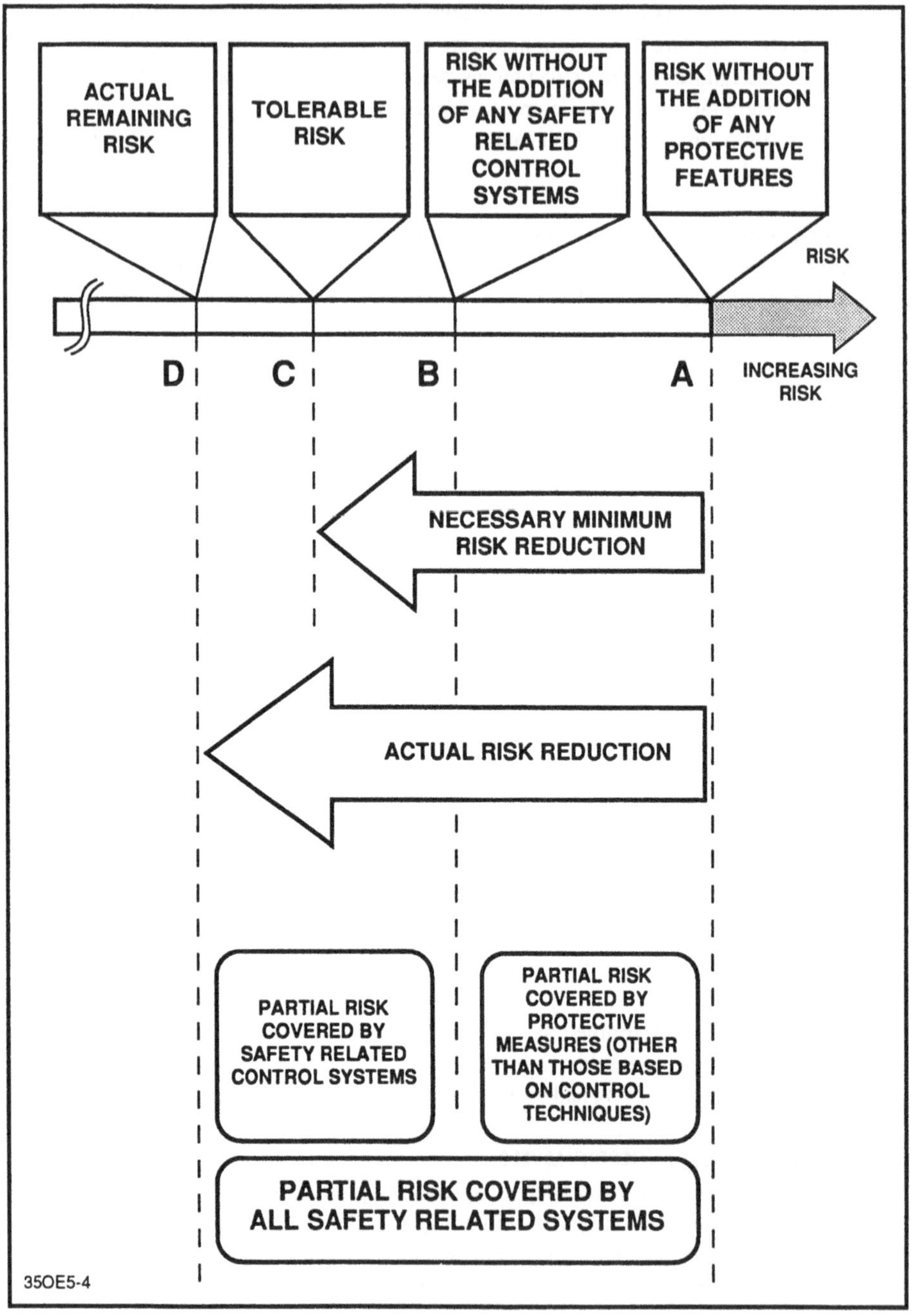

35OE5-4

281

Figure 5   Risk and Safety Integrity Concepts

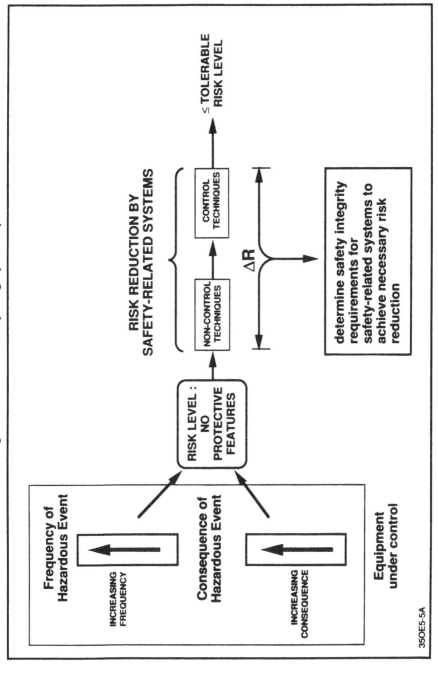

35OE5-5A

Also, it should be noted that for both systems and software, Integrity Level 1 is the lowest level and Integrity Level 4 is the highest. (This is the reverse notation to what was proposed in the draft International Standards. The new notation was adopted subsequent to the consultation carried out). The IEC draft standard dealing with safety- related software[7] has 4 software Integrity Levels.

The general concept for system and software Integrity Levels is shown in Figure 6. A simplified example to illustrate the concept of system and software Integrity Levels is outlined below. It is assumed that the tolerable risk level has been specified and it has been ascertained the necessary risk reduction (ΔR), to achieve the tolerable risk levels would be met if the combination of the individual safety-related systems gave an Integrity Level of 3 (that is, the Integrity Level for the total - configuration of safety-related systems is 3). Each safety related system carries out the required safety functions independently of the other. One of the individual systems is a hard wired system and the other individual system is a programmable electronic system (PES). It can be seen from Figure 7 how the Integrity Level of 3 for the total configuration of safety-related systems is derived from the 2 individual safety-related systems both of which one have an Integrity Level of 2.

Note: *The system rules for adding together individual systems of a specified integrity level into an overall integrity level and how the software integrity levels would be established have yet to be finalised.*

9. SAFETY LIFECYCLE
During the development of the draft Standard the Working Group realized that certain processes were necessary in engineering any PES-based safety-related system. These processes fall in a natural chronological sequence, similar to many conventional design projects:

- analysis

- specification

- design & implementation

- validation

- maintenance

The unique processes associated with developing and proving a PES-based safety- related system can be extracted from this design sequence and form the basis for a Safety Lifecycle. The advantages of the Safety Lifecycle concept are:

Figure 6   Risk reduction and Safety Integrity Levels : Overall Framework

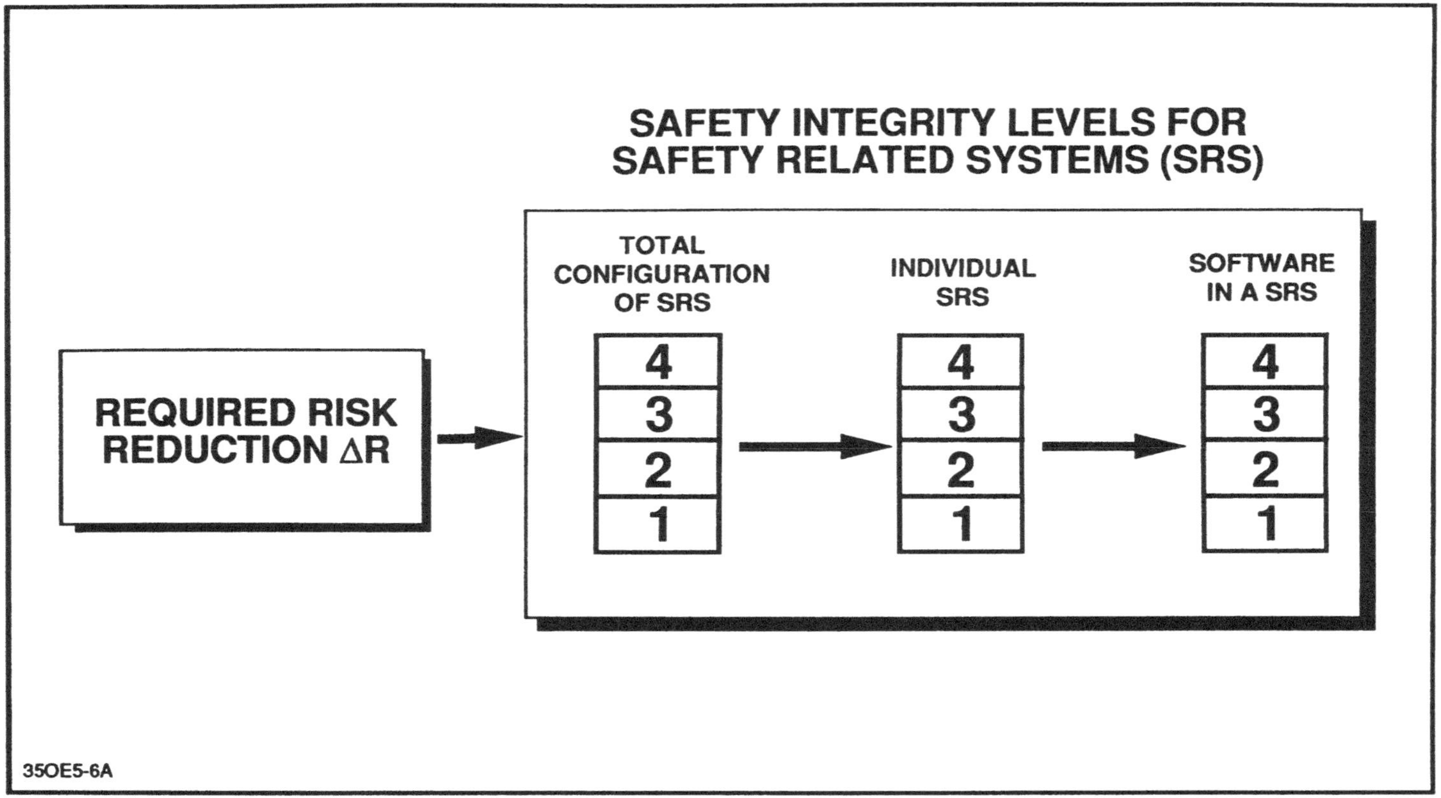

284

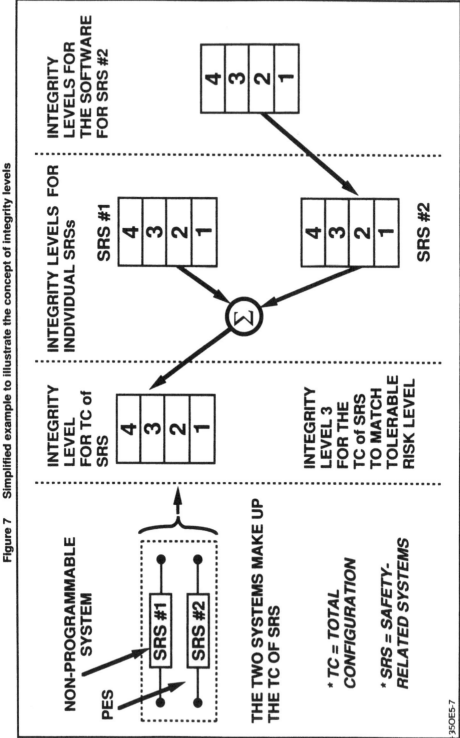

Figure 7 Simplified example to illustrate the concept of integrity levels

35OE5-7

- It provides a framework for explicitly identifying all of the phases required in the development of a safety-related system, and thus provides a framework for the draft standard itself.

- It relates the development phases, both chronologically and structurally.

- It provides a structure for analyses documents, design documents and test documents which must be produced.

- It is congruous with the development lifecycle of conventional, "non-safety-related" equipment. Therefore, much of the existing work on design quality assurance is directly applicable.

The present form of the Safety Lifecycle is shown in Figure 8.

9.1 Hazard analysis and risk assessment

The hazard analysis is a structured analysis, the extent of which depends on the seriousness of the risk in question and the complexity of the systems involved. It may be performed using any of a variety of accepted tools, including:

- fault tree analysis (FTA);

- failure mode, effect and criticality analysis (FMECA);

- hazard and operability studies (HAZOP).

The objective of the hazard analysis is to identify all the hazards together with the events which could give rise to them.

The risk assessment is an analysis which establishes the risk level of the EUC (without the addition of any safety-related systems) and determines the risk reduction (ΔR) necessary to achieve the tolerable risk levels. The risk assessment may be qualitative or quantitative, depending on the risk level. The objective of the risk assessment is to determine if the risk level is lower or equal to the tolerable risk level.

9.2 Safety requirements specification

The **Safety Requirements Specification** is divided into the:

- Functional Requirement Specification

- Safety Integrity Requirements Specification.

Figure 8 Safety Lifecycle Model

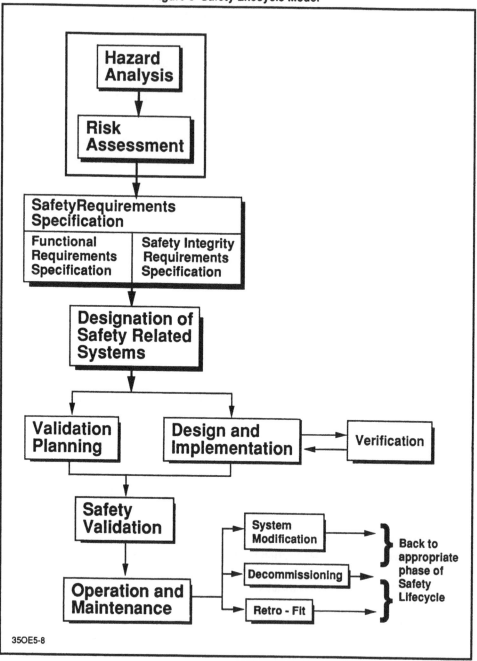

35OE5-8

The **Functional Requirements Specification** establishes the **safety functions** which must be carried out by the safety-related control systems. For example, a functional safety requirement for a thermal shutdown might be:

- if TEMP > HIGH SET then OPEN VENT and CLOSE FEED.

The **Safety-Integrity Requirements Specification** establishes the requirements for achieving the **safety-integrity** necessary for performing the safety functions. The Safety-Integrity Requirements are intended to establish a degree of certainty with which the safety functions will be executed.

The Safety-Integrity Requirements Specification establishes the requirements for the three system safety elements:

- **Configuration**

- **Safety-Related Hardware Reliability**

- **Systematic Integrity.**

9.3 Designation of the Safety Related Systems

Three levels of "system" are defined in the draft Standard:

- **Total configuration of systems:** comprises all of the elements that make up the hierarchy of systems for control, protection and monitoring of the EUC.

- **Total configuration of safety-related systems:** comprises all the safety-related systems which, together, achieve the necessary level of safety integrity.

- **Safety-related system:** comprises a system that achieves, on its own or with other safety-related systems, the necessary level of safety integrity for the required safety functions.

It is important to separate, identify and designate those systems which are safety-related, since they are responsible for achieving the overall safety and they are the systems to which the remainder of the Safety Lifecycle applies. Figures 4 & 5 shows the concept of the safety-related systems and their role in reducing risk to the tolerable risk level.

9.4. Design and Implementation

Design and implementation of a safety-related PES is similar to that of a conventional system. However, the draft Standard calls out two specific requirements of the design process which are specifically required for safety-related PESs:

- Design and implementation shall be carried out in accordance with an established quality standard such as ISO 9001.

- Safety Verification must be carried out in conjunction with the design and imple mentation process.

Safety verification is the comparison at each stage of the design and implementation process that there is a faithful translation from one stage into the next (beginning with the Safety Requirements Specification).

9.5 Safety Validation

Validation is the process of determining the level of conformance of the final operational system to the original Safety Requirements Specification. Validation consists of executing comprehensive tests and applying mathematical and systematic analyses.

9.6 Operation and Maintenance

Operation and maintenance requirements in the standard consist of monitoring safety performance and maintaining controls over design changes.

9.7. System Modification and Decommissioning

These areas are not addressed in the present standard and are under consideration for future work.

10. WAY AHEAD

The next phase in developing the draft Standard is to finalise Part 1 'General Requirements' and to begin drafting future parts. Finalisation of Part 1 is essential in order that those responsible for developing application - specific standards can begin to incorporate the requirements set out in Part 1 into their own standards (see Figure 2). After receiving comments from the National Standards Committees, it is hoped that Part 1 can be finalised taking into account the comments received. It is envisaged that at least one further draft will be distributed for comment before Part 1 becomes a published International Standard.The key issues remaining to be resolved, with respect to Part 1, include:

- **Integrity Levels:** There is a need to finalise the number of system Integrity Levels.

- **Summation of Integrity Levels :** The"rules" for summating the Integrity Levels for individual safety-related systems into the Integrity Level for the total configuration of safety-related systems have to be developed.

- **Configuration System Element :** The requirements for this system element, which is intended to add "robustness" tothe overall system design, has to be finalised.

- **Safety Lifecycle :** Is the order of the key elements that need to be addressed chrono logically correct, or are modifications required relating to the Safety Requirements Specification and the Designation of Safety- Related Systems?

• **Systems - Software Interface** : The draft standard on safety-related software must have clear interfaces with the draft standard for safety-related systems. Further consideration is required to ensure this requirement has been met.

Finally, it is vital to the future effectiveness of Part 1 that feedback from the target audience be obtained (ie. those involved in the development of application - specific standards). This feedback is necessary to make the final Standard sufficiently robust to cover the wide variety of systems with which it will have to deal, and workable, and of direct practical benefit, to standards makers and others who need to use the Standard. There are many challenging issues yet to be solved but the opportunity exists of developing a standards framework relevant to all application sectors which will have important safety and economic advantages. Such an opportunity should not be missed.

11. REFERENCES

1. Programmable electronic systems in safety related applications:
   "1 An introductory guide."
     ISBN 011 8839136
   "2 General technical guidelines."
     ISBN 011 8839063.
Both documents available from HMSO, PO Box 276, London SW8 5DT. (Telephone orders: 01-873-9090; Fax 01-873-8463).

2. DIN V VDE 0801: Principles for computers in safety-related systems (Grundsatze fur Rechner in System mit Sicherheitsaufgaben).

3. Principles for computers in safety-related systems; K Meffert; ACOS Workshop on Safety-Related Control Systems (An International Standards Workshop on Programmable Electronic Systems for use in Safety-Related Systems); IEE, Savoy Place, London; 8-9 March 1990. (Proceedings available from Publication Sales, PO Box 96, Stevenage, Herts SG1 2SD, UK, Price £15 Europe and rest of world; $30 USA; Telephone 0438-313311; Fax 0438-742792).

4. The application of programmable electronics systems to the control of machinery: The scene in Europe; B J Clark, G R Ward. ACOS Workshop; (see Ref 3).

5. Guidelines for safe automation of chemical processes; T Lagana ACOS Workshop; (see Ref 3).

6. IEC draft standard "Functional safety of programmable electronic systems: Generic aspects: Part 1: (IEC reference "65A (Secretariat) 96").

7. IEC draft standard "Software for computers in the application of industrial safety-related systems" (IEC reference "65A (Secretariat) 94").

FIRE DAMPERS - TAKING STOCK

G. Swinton: Man.Director - S.G.L. Systems, Ltd.

SUMMARY

Fire Dampers are a device for preventing the passage of smoke and flame through ventilation ducts and bulkhead penetrations.

These devices are subject to rules by Legislative Authorities.

This paper is written for personnel specifically involved in the Safety, Loss Prevention and Heating and Ventilation disciplines which interfaces with electrical, instrumentation and architechtural departments.

GENERAL

Since the early seventies, when the North Sea became more than a place to gather fresh fish and a route to the Continent, we have seen the construction and erection of all sizes of steel and concrete structures supporting a lifestyle and workforce that Jules Verne would have admired.

These habitats have advanced considerably over the years, from wooden huts to the standard of luxury hotels. In order to keep the workforce happy, living accommodation is required to be, among other things, comfortable, safe and with good occupational facilities.

It is the safety aspect that we look at and in particular, the control of fire hazards of flame and smoke.

In all areas of offshore structures ventilation fans are installed to supply fresh air for environment or combustion purposes. Whenever a ventilation duct passes through a rated fire wall either, supply or extract, a fire damper must be fitted at the penetration point, in order to close off and segregate the particular area, whether galley, radio room or living area.

The first fire dampers installed in the offshore structures took the form of simple hinged flaps held with a chain and a fusible link. On the fusible link melting at a set temperature the flap dropped under gravity and closed off the duct area.

Later development incorporated a concertina type blade still held by chains and fusible links. These units were a copy of similar units used in the indus-trial buildings and construction industry.

Current contract specifications and statutory instrument S.I.611, enforced by the fourth edition of D.O.E. Guidance Notes - indicate guidelines for latest construction - Para 47.5.4. Fire/Gas and Watertight Dampers (5):-

"Selection of appropriately robust equipment certified for the particular application should be made. These dampers should be capable of withstanding temperatures and pressure for the rating of the barrier, and also of responding automatically to alarms, with provision for local and remote operation and indication to meet the emergency shutdown logic of the installation.

Dampers should have the minimum leakage practicable for the particular application involved.

Dampers required to control supply of combustion air to engine (rig savers) should be of the same robust construction as fire/gas dampers. Controls and damper actuators should be inherently non-sparking and appropriate to area classification. Actuators should be capable of rapidly closing and opening the dampers against airflow pressure within the duct. Controls should be arranged to 'fail safe' in the event of loss of power or break-ing of fusible or frangible link(s). These links should be positioned where they will be able to detect fire or over temperature; the arrange-ment of links required to provide adequate protection should be analysed for each application to ensure that the barrier is correctly protected."

It is known on some current projects that the flimsy concertina type dampers are still being used - mainly for economic reasons. Policing of these projects must be rigorous and detailed if we are to maintain the required standards.

EXISTING STANDARDS

Due to the initial rapid growth of the Offshore Industry, there was not; and there is still - no unfied fire test standard for the testing of offshore plat-form fire dampers.

The only criteria available is a combination of rules used in marine and construc-tion industries, that is, Safety of Life at Sea (SOLAS) and BS476. It is now considered that these regulations are no longer compatible with the offshore requirements and testing of fire dampers is currently under review in ISO and BSI Committees.

GENERAL REQUIREMENTS

The basic requirements for passive fire protection of offshore structures are similar to both United Kingdom and Norwegian waters. They are based upon the requirements used to ensure safety on ships, adpated to the specialised needs of a offshore hydrocarbon producing installation. The basic requirements regarding passive fire protection are laid down by the 1974 SOLAS convention, the 1978 SOLAS Protocol and the 1981 and 1983 SOLAS Amendments. Abstracts from these are also used by the Acceptance Authorities in their regulations for offshore units.

The SOLAS convention defines the standard cellulosic fire test by temperature time curve. 'A' class divisions shall be of steel or equivalent material, suitably stiffened and be constructed so as to prevent the passage of smoke and flame to the end of the one hour standard fire test which is defined by a temperature-time curve.

Insulation requirements are laid down for different 'A' Class divisions and different time periods, A-60 to A-0, 'B' class divisions are required to prevent the passage of flame for the first one half hour of the standard fire test and to maintain insulation requirements for given time period B-15 and B-0.

Intergovernmental Maritime Organisation IMO resolutions define test procedures enabling classification of materials as non-combustible and the standard fire test on divisions.

The Department of Energy Hydrocarbon Fire Resistance Test for Elements of Construction for Offshore Installations - Test Specification, Issue 1, January 1990 (2), now defines test specifications for primary and secondary elements of construction.

Offshore Installations: Guidance on Design, Construction and Certification: Fourth Edition 1990: Adddresses this subject (5).

"Section 47 - Para 13.3.   Fire Protection Materials and Their Testing".

Requirements for divisions formed by bulkheads, decks, ceilings and linings should conform to the as amended 1974 SOLAS convention and are based on the A-60, and B-15 class divisions. Specific requirements cover these applications and include door, penetrations, draught stops, internal stairways and control stations. Requirements for boundaries and ventilation separating machinery space, storerooms, galley accommodation etc., and related again to the standard class divisions.

Future changes to the regulations following the Piper Alpha disaster are anticipated. The Department of Energy discussion document PEA 584/411/7 is proposing formal safety assessment of each installation which will be used to derive the necessary levels of protection and performance criteria in the various areas. Protection from the affects of explosion will be required and there will be an increased need for H-120 class protection.

Current Acceptance Criteria

Protection against fire can be divided into two broad categories, to provide for the safety of personnel and to provide for the protection of the installation.

Legislation applicable to offshore applications is found in Acts, Regulations and Decrees both in the United Kingdom and Norway. These cover Mineral Workings (Offshore Installations), Petroleum and Submarine Pipelines, Exploration and Exploitation of Petroleum, Fixed Indstallations and Mobile Installations.

In the United Kingdom certification as 'fit for purpose' in accordance with
Statutory Instrument 1974 No. 289 The Offshore Installations (Construction and
Survey) Regulations 1974 may be undertaken by the following authorised
organisations:-

American Bureau of Shipping

Bureau Veritas

Det Norske Veritas

Germanischer Lloyd

Offshore Certification Bureau

Lloyds Register of Shipping

Practical implementation of the Norwegian acts, regulations and guidelines for
fixed offshore structures is co-ordinated by Norwegian Petroleum Directorate
(NPD). Supervising activity being delegated to such bodies as Maritime
Directorate and Coastale Directorate. Control on mobile drilling platforms
is co-ordinated through Norwegian Maritime Directorate (NMD).

SPECIFIC REQUIREMENTS FOR STRUCTURAL FIRE PROTECTION

Department of Energy

U.K. Department of Energy guidance notes cover fire protection for accommodation
and control spaces and equipment.

Lloyds Register of Shipping

Lloyds publish separate rules and regulations which apply to mobile offshore
units and fixed offshore installations. Whilst the regulations contain a chapter
that specifically deals with manufacture, testing and certification, protection
of a structure against fire is based on the revised requirements of the 1974
SOLAS convention. The classification is further broken down to include a C
class division constructed of approved non-combustible materials, but needing
to meet much lesser fire requirements.

NPD Regulations

Structural fire protection by active means is to be arranged so as to prevent the
fire spreading to ther areas and to reduce the consequences of explosion. Class
A and B fire walls similar to those laid down by Department of Energy are required,
but in addition H class fire walls are defined.

DAMPER RELIABILITY

At this point it would be reasonable to discuss damper reliability. In a study carried out by the Safety and Reliability Directorate UKAEA 1988. Mainly based on Offshore industrial equipment (1).

"Findings from site Surveys

Two types of problems were often encountered on sites - seizure of mechanism causing failure to close or open, or spurious closure due to fusible link failure. Detailed records of damper maintenance had generally not been kept by sites so it was difficult to establish the reason for failure. Seizure of dampers was believed to be caused by either distortion (twisting) during installation or bearing failure; link failure was believed to be a combination of creep, high ambient temperature or vibration. Enquiries made within the offshore industry highlighted problems many years ago with curtain type dampers, but more recently with the drying-out of oil impregnated phospher bronze bearings. Obviously components made of mild steel created problems offshore due to corrosion.

On a more general point, dampers for offshore are subject to high levels of quality assurance which are sometimes not applied in normal industry. The performance specifications for offshore are usually more restrictive than normal industry, which also tends to result in more reliable dampers.

However, lack of regular maintenance, no matter how good is the damper, results in unreliable operation."

It was interesting to note that "the recommend features" of Fire Dampers, in this study, were based on a typical Offshore Unit.

Recommended features for a Fire Damper (1)

Fire Dampers whose successful operation is important in terms of life safety and loss prevention need to be reliable. The following list of design features should assist toward this aim:-

a)   The damper including its operating mechanisms should be of suitable stainless steel.

b)   The thickness of damper casing and flanges should be a minimum of 3.2mm, or the same as the ductwork flanges if these are thicker.

c)   Damper blades should be a minimum of 3.2mm thick.

d)   Blade shafts should be at least 19mm diameter.

e)   The damper control mechanism should have as few a number of moving parts as practicable.

f)   Damper bearings should be easily accessible from outside the casing and protected from the ambient environment.

g)   Bearings should be of nickel steel or similar, eg carbon but oil free.

h)   A positive indication of damper position should be given, preferably by use of poppet microswitch.

i)   Blades should not be bolted to shafts, (vibration can cause problems).

j)   Thermal insulation should be provided between damper linkages and actuator control box.

k)   Blades, bearings, position indicators and control mechanism, should be easy and quickly replaceable without disturbing ductwork.

l)   Where a temperature sensitive closure element is provided, the design should allow the interchanging of frangible bulbs for fusible links should the links prove to be a source of spurious operation.

Discussion

The successful operation of fire dampers is important both in terms of life safety and loss prevention. Current fire test standards do not consider installed reliability; neither explicitly do relevant Acts and Regulations. However, test standards for fire protection equipment like detection systems, do address reliability by way of various environmental tests and some also specify minimum reliability requirements. Designers and Operators should be aware of these facts and consider damper reliability more carefully.

Reliability engineering experience of equipment similar to damper systems, suggests that a well designed and properly maintained damper, should achieve a probability of failure on demand - with quarterly testing - of 1×10^{-2} or slightly better. For a poorly designed damper with inadequate maintenance, the failure rate may be 0.5 per demand or worse. If for safety reasons values better than 1×10^{-2} are necessary, then two dampers in series, one either side of a fire barrier, would be required.

APPLICATION

The product, however, must meet the demanding requirements of the Oil Industry and in doing so lead the world in the development and sophistication of fire dampers and their associated and sometimes complicated control systems.

It is essential to take stock for the future, to meet the onslaught of regulatory bodies and to keep in front of proposed legislation.

As new and revised rules appear it is necessary to ensure that the existing platform systems are up to date.

To this end, the SAFEDAMP free area fire damper was developed, giving fully automatic opening and closing, remote resettability, and incorporating low power, low weight and maintenance accessibility. This damper can be installed in existing systems without the requirements of upgrading fan sizes and subsequently increasing the power sources. The free area fire damper (designated SFD611/2) has been successfully used on many offshore installations for both refurbishment and new construction projects.

Should a fire damper also be used to prevent the ingress of gas? On many occasions designers have required a combined fire and gas damper. By virtue of the current testing criteria. Dampers are allowed to have gaps of such a size that, dependent on duct area, any amount of gas can penetrate.

While the designers requirement can be met by inserting sealing material to cover these gaps it cannot be the correct solution. If a tight gas seal is required then a damper specifically designed for such a function should be installed in series with a fire damper.

There is no criteria available from the regulatory and certifying authorities as to the permitted levels of gas ingress through a closed damper.

STATE OF ART FIRE DAMPERS

The Concept of Safety with Fire Dampers

If the spread of flame and smoke from an outbreak of fire is to be prevented effectively, it is essential for the fire to be contained behind the barriers surrounding the outbreak.

This means that all openings through the barriers must be sealed. Motorised fire dampers are used for this purpose in air conditioning systems. They are triggered either directly by smoke detectors or from a central fire control station.

In the event of an outbreak of fire, all fire dampers in the zone of the outbreak, close simultaneously when triggered by a fire or smoke detector. Since the fire detectors and smoke detectors are grouped according to particular fire zones, their signals can be processed at the central fire control station to provide control commands for the fire dampers. With this type of arrangement it is also possible to re-open fire dampers for somoke extraction purposes under certain circumstances.

Motorised fire dampers are ideal for allowing an objective check of the proper functioning of the dampers. The check is performed by operating all the dampers in a particular zone, and position feedback (either individual or common) shows whether complete sealing has been achieved. A fault alarm is given automatically at the central fire control station if there is any discrepancy between the control command and the position feedback.

The control commands to the fire dampers and their position feedback signals are grouped in a control module according to their assignment to particular fire zones.

This permits easy connection to higher-level supervisory systems since the number of control commands and feed-back signals needed can be greatly reduced.

Although it might be linked to a higher level supervisory system, the control module still allows the local checking of individual dampers because the position feedback signals, and any fault alarms, are indicated seperately.

There is also a facility for local electrical/pneumatic operation of installed fire dampers in order to check their proper functioning. Electrical installation requires minimum cores and cables thanks to the grouping of the wiring for the individual fire dampers in a single connection box and the processing of the position feedback signals in a single control module.

The use of standard control commands and suitable position feedback signals in the control module makes it the ideal electrical interface for higher-level supervisory systems.

HYDROCARBON FIRE TESTS

As previously mentioned there is no uniform fire testing procedure for fire dampers. However, this problem has been addressed relative to Hydrocarbon Fire Tests.

The fourth edition of Department of Energy Guidance Notes define Hydrocarbon boundaries and criteria for testing.

Test Specification Issue 1 - January 1990 (2). Fire Research Station, gives a time/temperature curve. The Test procedure document indicates procedures for all penetration elements, doors, windows etc. However, it clearly states in Chapter 2.8 - Dampers - that there is no ISO or other reference standard for this chapter.

As generally it is the case it requires the Manufacturers to be forerunners in these situations.

A test based on this document has been arranged at Warrington Research Centre. If successful then possibly we will have a 'Basis for a Standard'.

STANDARDISATION OF OPERATOR REQUIREMENTS FOR FIRE DAMPERS

In order to provide a product which can be produced to meet the fire test requirements and also meet all the varying opinions of constructional and control operating systems is virtually impossible.

Various manufacturers have their own design of product which has been certified by the Regulatory Authorities and therefore meets the test parameters and is suitable for installation in the relevant locations.

However, contractor written specifications which vary from contractor to contractor, can be very specific in their requirements, including constructional detail, which can eliminate some manufacturers products from tender lists. This hardly seems fair when the development costs involved can be substantial.

What is required is a basic requirement for an approved product with control system preferences. This would result in cost savings both in contractor design time and the resultant end product.

It must be noted that design deviations from the tested samples would invalidate the approval certification. Quality audits can be carried out by the Regulating Authorities to ensure compliance.

REFERENCES:-

1)   The Reliability, or Unreliability of Fire Dampers.

     M. Finucane,
     Safety & Reliability Directorate - U.K.A.E.A. Warrington.

2)   The Hydrocarbon Fire Resistance Test for Elements of Construction for
     Offshore Installations.
     Test Specifications ' Procedures Issue 1 - January 1990.   FRS 14-84.

3)   ISD/TC92/SC2/WG4 - Fire Resisting Dampers
     Part 1 - Method of Test.

4)   IMO Resolution A517 - Part C.

5)   Offshore Installations -

     Guidance on Design, Construction and Certification (Fourth Edition).

6)   BS 476.

The Author also Acknowledges discussions and comments from+-

    Shell UK Ltd.

    Mobil

    British Petroleum.

    Brown & Root Vickers, Ltd.

THE APPROACH TO TRANSIENT PROBLEMS

IN FIREWATER SYSTEMS

PETER MILES
GENERAL MANAGER ENGINEERING
CHUBB FIRE ENGINEERING, UK

NOMENCLATURE

DELTA H   =   Change in meters head

DELTA V   =   Change in velocity mtrs/sec

C   =   Acoustic velocity mtrs/sec

g   =   Acceleration of gravity

W   =   Mass

K   =   Bulk modulus

P   =   Pressure

d   =   Pipe diameter

t   =   Pipe wall thickness

E   =   Youngs modulus

e   =   Effective density

ε =   Free bubble content

Summary

Surge is potentially a very damaging force which is difficult to contain. It is better to design it out than incur the significant additional costs to contain the pressures. Also subsequent system modifications could require additional expensive strengthening.

1.0 DEFINITION OF SURGE

Many people have difficulty in defining what is surge. In simple terms it is the short term pressure changes resulting from short or rapid velocity change.
When speaking about surge we are discussing upset conditions which occur separately from the steady running of the pipe.
In considering the phenomena it is very similar to the transmission of sound, human speech being complex surge patterns transmitted through the air. It must not be confused with shock waves, which cannot occur in liquid flow conditions.

Fig.1

$$DELTA \quad H = \frac{C\ DELTA\ V}{g}$$

The basic physics of this phenomena is summed up in this simple equation which shows that the pressure change, **Delta H** is caused by a momentary step in velocity, and is proportional to the velocity change Delta V; the constants of proportionality are "C" the speed of sound in that liquid in that particular piping system at that moment in time, divided by the gravitational constant. For a flow stoppage occurring within a relatively short period, Delta V represents the change from line velocity to zero, and hence this equation can be used directly. But when the flow change is longer, then other factors constantly modify the velocity decay and in turn the pressure change.

Included in these effects are the nature of a valve that is closing or opening, the way in which a pump starts up and runs down, friction, the length of the line involved etc. Thus in real life the surge pressure change is less than the value suggested by the equation other than for extremely rapidly occurring events. This modification is important.
In order to define the size of any engineering problems arising from surge the surge pressure must be defined. Thus, by over simplifying surge calculations surge pressures are overestimated, leading to badly spent money and in exceptional circumstances, a complete miss appraisal of the surge problem.

The previous formula showed that the change in surge pressure is proportional to the velocity change. It also showed that it was proportional to the speed of sound in the liquid.

This equation shows how the speed of sound value can be derived.

Fig. 2

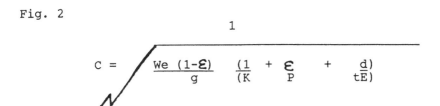

$$C = \sqrt{\dfrac{1}{\dfrac{We\,(1-\mathcal{E})}{g} \left(\dfrac{1}{K} + \dfrac{\mathcal{E}}{P} + \dfrac{d}{tE}\right)}}$$

The first term, concerning the Effective Density is modified by the presence of any bubbles in the flow (the proportion being here represented by the Greek sign '$\mathcal{E}$' Epsilon).
The second, the compressibility of the liquid is particularly important. Many people prefer to think that liquids are not compressible, but in fact some petrocarbon products are particularly compressible. When considering the next term, the free bubble content, we will see that the compressibility effect of the pure liquid can easily be swamped in the presence of even a very small percentage of bubbles, as you will appreciate that the compressibility of air is many thousand times lower than that of liquid. The p (pressure) beneath the Epsilon sign allows for the modification of the bubble percentage with the particular pressure that is applied at a given instant at a given point on a given pipeline. The final term, covers the distensibility of the pipe represented by the diameter of the pipe divided by the product of the wall thickness and the elasticity of the pipe wall material.
Summarising, the most important factor in deciding the celerity or speed of sound is the free bubble content. The next most important term is the pipe distension (unless we are looking at particularly thick walled pipes in very strong material). The compressibility of the liquid, in the absence of gas is also highly dominant. So we have a complex interplay between groups of numbers which causes the speed of sound to change greatly.

Example:

The speed of sound in the human arterial system can be as low as 15ft per second. Whereas crude oil in a thick walled steel pipe can have a speed of sound as high as 3500ft per second. Water, in a rock tunnel could well have a speed of sound of 4200ft per second. The presence of gas or its evolution due to pressure drops in surge events, can vary the speed of sound in an oil pipeline from 3500ft per second to 400ft per second.

2.0   COMPUTERS

It is therefore not surprising that although these effects have been known for some seventy years, precision calculation and prediction of surge pressures has only been realistically possible over the last nineteen or twenty years. But the advent of large high speed computers and sophisticated programming techniques permit us to take into account not only the many variables affecting surge pressure changes in a pipe but also to write computer programmes matching the variables associated with the valves, control system, pumps, pump drives, branches and even the characteristics of computer surveillance of sophisticated pipelines.

2.1   By taking into account all these variables and using the appropriate computing power, pressure changes can be calculated with very high precision. This means that all the operating problems can be determined well before the design of the system is finalised. Alternatively, on an existing system, by using the computer as a precision simulator, tests of events can be made which would be unsafe to carry out on the real life system. The quality possible from good surge analysis is such that there is no loss in engineering or economic reliability.

3.0   TOOLS TO DO THE JOB

Do not expect to be able to do good surge analyses on your desk top calculator. Not only does the equipment require specialised personnel to carry out these analyses but suitable computer hardware is required in order to carry out such calculations economically. It can be demonstrated that simplification of the calculation processes for use on a simple computer, rapidly increases the possibility of random error in the results for even minor simplification of the process.

4.0   ENGINEERING CONSEQUENCES OF SURGE

Returning more to the engineering consequences, a typical pressure against time trace of a valve closure is shown in Fig. No.3. As any valve closure or pump start will cause a change in velocity, a surge pressure rise upstream of it will ensue. The total surge effect occurs by changing the initial line velocity at the time the valve is only just beginning to close, to the peak pressure which occurs at the moment of valve closure.

4.1   The very great rise in surge pressure is not necessarily the magnitude of the surge problem. If the surge pressure rise does not exceed the pipe rating then there is no surge problem and there is no requirement for any further action by the engineer. Please note that the shape of the surge pressure rise during valve closure will alter. Very little pressure rise occurs over the majority of the stroking time of a valve but very rapid pressure rise occurs during the last stages of closure.

5.0   VALVE CLOSING CHARACTERISTICS

The surge pressure rise due to closure of any valve is related to its flow characteristic Cv, and is in fact proportional to 1/Cv2. The severity of surge pressure rise during closure eases with valve type. Strictly speaking, this statement is only true when the time of closure of the valve is several times the natural surge periodicity of the piping system being considered. If the time of one surge period is equal to the length of valve closure then the surge pressure rise is in total almost the same for all types of valve. However, if the natural periodicity of the system we are looking at is shorter than the valve stroking time so that only the last quarter of the valves closure is relevant in surge terms, then the difference between the types of valves is significant. Thus if we have a surge problem we can ease that problem by changing the type of valve or its length of closure. How effective these measures will be depends upon the natural periodicity of the piping system in which the valve is installed.

6.0   TYPICAL CAUSES OF SURGE IN A PIPE NETWORK

The most common causes of surge arise from pump start up, pump failure or valve closure against flow. There are also secondary surge problems associated with resonance, control valve interaction, the creation and collapse of vapour or gas pockets etc.

Figure No 3

TYPICAL PRESSURE/TIME SURGE PROFILE

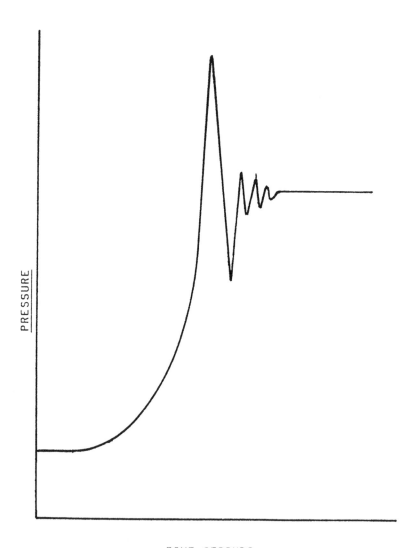

PRESSURE

TIME SECONDS

6.1   Taking a typical piping system, action at various points throughout can potentially give rise to surge problems. This simplified to an incoming pipeline and tank farm, with tanker loading facilities. Flow in the main line could be stopped by block valve closure at the terminal or shutting isolating valves in the tank farm during changeover from tanks. This is particularly important as the pressure rating of the tank farm is usually well below that of the pipeline and hence extremely serious situations could develop. Similarly, shutdown either on the jetty or on board the tanker will cause an equally serious hazard, if the tanker is being loaded either directly from the pipeline, or from the tank farm. Further surge problems can occur during normal tanker loading operation when flow from the tank farm will be pumped to the vessel. Trip out of the pumps will cause a rapid surge pressure rise upstream of the pumps. Alternatively, if the tanker is discharging to the tank farm then valve closure at the jetty, at the pump bypass valve or closure of any of the valves leading to the storage tank will similarly give rise to surge problems. Thus, it is prudent to inspect all routes of flow and operating conditions and seriously consider where the surge hazards can occur and where necessary determine the magnitude of the pressure problem to decide whether there is an engineering problem.

6.2   Having determined the severity of the surge pressure conditions and identified our engineering problems we must now consider the possible solutions. In some systems moderate surge problems can be eased by reducing the rate of valve action or by choosing a different type of valve. Alternatively, the traditional solution is to purchase excessively thick pipes which can contain the surge pressures; but due to the cost of construction, this is no longer an economic solution. Clearly, as surge pressure rises are proportional to the velocity in the system the surge problems become less as the flow rates decreases.

7.0   SURGE IN FIREWATER SYSTEM

The first element of the analysis is to look at the priming of the dry pipe underneath the check valve from the fire pump into the ringmain. This study includes diesel run up times, fire pump curve and priming times of the dry riser. This will establish a datum point to work from of a worse case condition of a fire pump starting against a closed valve system. From this datum point we can then move into the ringmain and see what effect that would have on the hydraulics of the closed circuit system. We have to take educated guesses at the air volume contents of trapped gas around the ringmain (which is forever present in various locations, typically up under the helideck). This study will have to be conducted with varying volumes to obtain a distribution curve and a worst case condition. It is important to establish this worst case condition to ensure that during the engineering phase that those circumstances are avoided wherever possible and designed out.

7.1 The next step is to introduce the activation of a deluge system which would allow the ringmain pressures to fall below the vapour pressures of the water at the head of the risers up under the helideck, (being the highest point on the ringmain). As the static head needs to be in excess of this height to maintain a water column, it is very important to establish the critical size of deluge take-up combined with fire pump start up that is going to give the worst case condition for surge generation in those risers, thereby resulting in ringmain instability. It may not be the biggest system or the smallest system that gives us the worst case, it could quite easily be one of the middle sized systems down on the cellar deck which generates the pressure decay characteristics when combined with fire pump start up, gives us the worst case condition for column collapse in the helideck risers. This basic data that would be generated during the design study would then indicate the areas of the design to be avoided and the timing of the closure of the various valves and response rates that we are looking for in the fire pump overboard dump system as it will become evident closing that valve too slowly can be just as critical as closing it too fast. We have to balance the charging of the ringmain with the pressure in the ringmain and the speed if the rising columns from the helideck level to obtain a smooth recharge up to steady state conditions.

8.0   ENGINEERING DATA

The engineering data necessary to conduct a front end study would be P & I Ds, piping isometrics fire pump characteristics, diesel driver characteristics, estimated deluge take-off requirements and sizes, sprinkler valve design specification, ringmain isolation check valve data sheets and an agreed air volume range to simulate the trapped gas around the ringmain. As you will see from the attached case study, the basic information would be readily available at this early stage of the contract. In addition to the direct fire fighting system information we would also need data pertaining to the seawater lift pumps and crossover arrangements; any pressure controlling valves, any relief valves on system, any relief or spill off valves or lines off of the ringmain used for water circulation or freeze protection. The front end study would be conducted in two phases - the first phase would be an engineering appraisal, the next phase would be moving into constructing a mathematical model of the basic ringmain leaving various parameters flexible and adjustable to be able to alter various valves as previously discussed, to establish worst case conditions. Having established the basic data, we can then move on to the hardware necessary to control the starting up of the pump to the overboard dump, its required sizing, water flows orifice plate for possible back pressure control sizing to keep the system as inexpensive as possible, as reliable as possible and the initial priming surges and resultant shock loadings within the piping code specification.

8.1   Having then established the sizing and timing of these valves, it would then be possible to determine whether there is any necessity to use anti-cavitation low noise valve trims. This part of the study is where consideration to reducing the capital outlay on the equipment, increases its reliability and simplify its maintainability with the aim to provide a totally water driven system requiring no external power source other than the water pressure generated from the fire pump itself. Having established a suitable engineering solution to the fire pump start up, we would then be reviewing the pump pressure controlling and relief system again sizing it as economically as possible.
Based on the space available with the allotted location area.

9.0   CONCLUSION

Many systems exist in service with many unexplained operational problems which have been identified by staff as just one of those things that happen after the pump starts or after the testing of a Deluge System. Even if a pipe ruptures under a Helideck or in the wet pipe sprinkler system it is put down to a weak link in the chain.
Piping codes allow for various design over pressures. They do not allow for the sort of pressure that can be generated by surge. Surge pressures can be 50 - 60 bars. The pipe may not rupture the first time but the material will be weakened and could rupture in 18 months to 2 years later during an incident. These problems can be engineered out right from the start.

10.0   ACKNOWLEDGEMENT

Hydraulic Analysis Limited, Leeds

CASE STUDY

DATA USED IN THE ANALYSIS

1.0   The purpose of this study is to consider the
general problems which occur in offshore platform fire
fighting system, particularly those composed of a ring
main and fed by dry risers from pumps situated down the
platform legs. As the study is not to be based on a
particular system, a typical mathematical model has to
be developed which is similar to many of the systems to
be found in the North Sea. The schematic shown in
figure 0.01 shows a simplified layout of the
mathematical model.

1.1   Three main fire pumps have been used,
nominally operating on a two duty/one standby basis.
These pumps feed seawater into the firewater ring main
which then feeds to hydrants and sprinkler/deluge
systems. The schematic layout shows the deluge system
used in the study connected through a valve to the ring
main.

1.2   It is assumed that the pressure within the
ring main is maintained at 9 barg by a jockey pump.
The first of the main firewater pumps is started if the
pressure in the ring main falls below 8 barg due to an
increased demand. The second pump is started if the
pressure fails to rise. The pumps are deep well
turbine pumps design to give a duty flow of 750
cu.m/hr. At this flow rate, the pressure in the ring
main downstream of the connection from the pumps would
be at average tide conditions of 0m.

1.3   The assumed elevation of the main pipework
are shown on figure 0.02. The important levels are:

Ring Main	30m a.o.d.
Sea Level	0m a.o.d.
Pumps	10m a.o.d.
Pump check valves	28m a.o.d.
High level Deluge	60m a.o.d.

The ring main is assumed to be constructed from 14 inch
diameter steel pipework in a 50m square, while the pump
risers are assumed to be 10 inch diameter. The feed
pipework to the hydrants and deluge system is assumed
to be 8 inch in diameter.

The pipe dimensions used are:

Nominal Diameter	Internal Diameter	Wall Thickness
8 inch	202.7 mm	8.18 mm
10 inch	254.5 mm	9.27 mm
14 inch	336.5 mm	9.53 mm

1.4   The performance curves for the pumps selected are shown on graph 0.03.

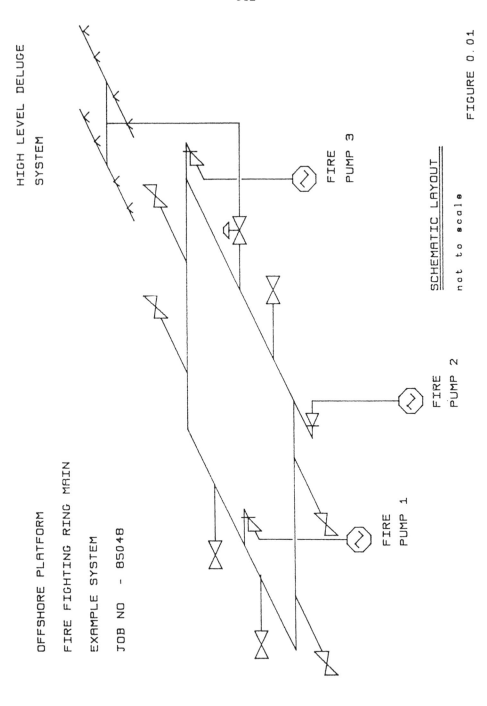

OFFSHORE PLATFORM

FIRE FIGHTING RING MAIN

EXAMPLE SYSTEM

JOB NO - 85048

HIGH LEVEL DELUGE
SYSTEM

FIRE
PUMP 3

FIRE
PUMP 2

FIRE
PUMP 1

SCHEMATIC LAYOUT
not to scale

FIGURE 0.01

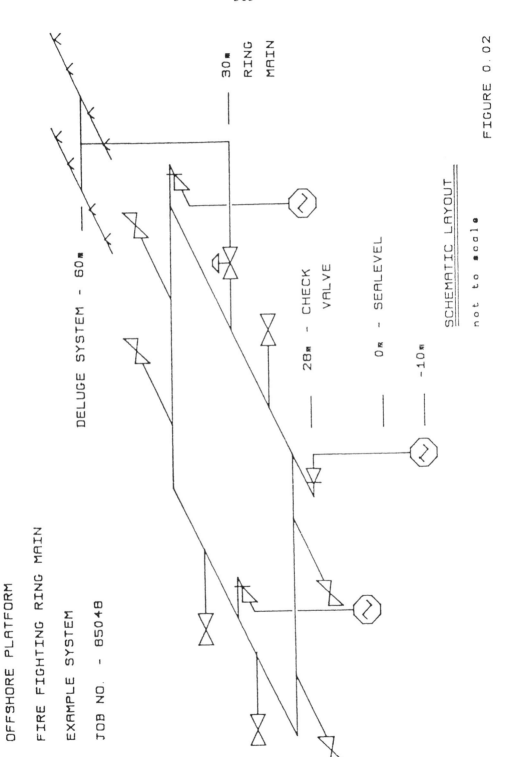

OFFSHORE PLATFORM

FIRE FIGHTING RING MAIN

EXAMPLE SYSTEM

JOB NO. - 85048

DELUGE SYSTEM - 60m

30m
RING
MAIN

28m - CHECK
VALVE

0m - SEALEVEL

-10m

SCHEMATIC LAYOUT

not to scale

FIGURE 0.02

314

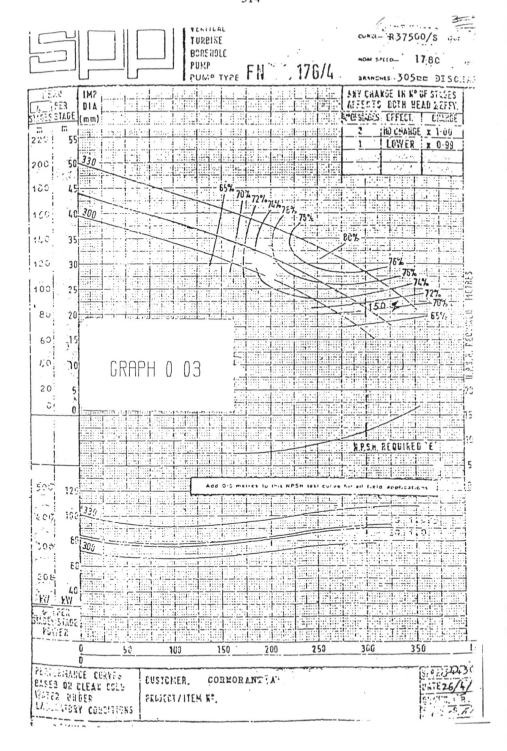

GRAPH 0 03

GRAPH NO. 1.01

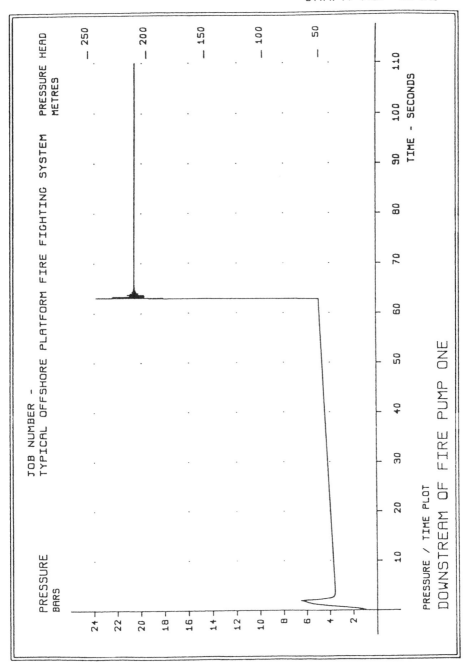

JOB NUMBER -
TYPICAL OFFSHORE PLATFORM FIRE FIGHTING SYSTEM

PRESSURE / TIME PLOT
DOWNSTREAM OF FIRE PUMP ONE

GRAPH NO. 1.02

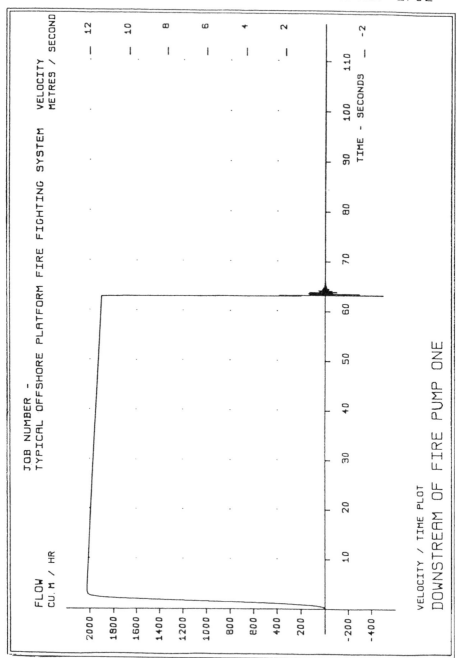

VELOCITY / TIME PLOT
DOWNSTREAM OF FIRE PUMP ONE

GRAPH NO. 1.03

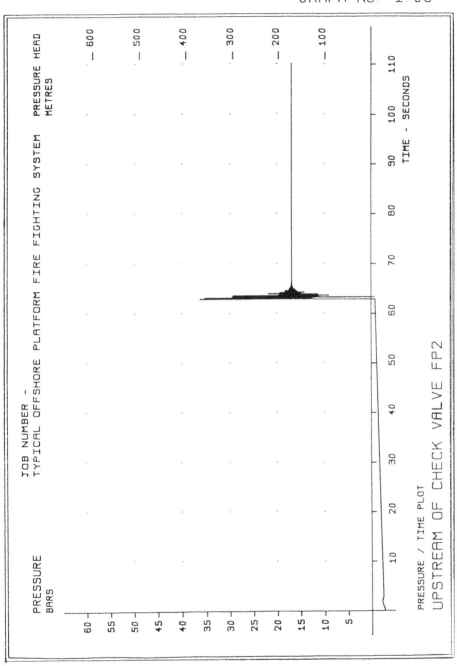

PRESSURE / TIME PLOT
UPSTREAM OF CHECK VALVE FP2

GRAPH NO. 1.04

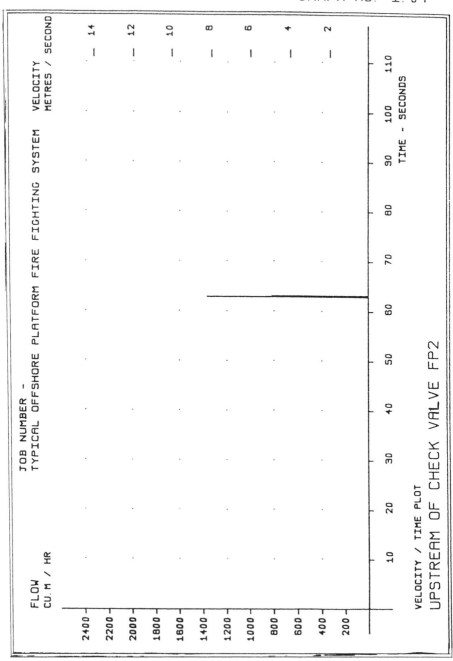

JOB NUMBER -
TYPICAL OFFSHORE PLATFORM FIRE FIGHTING SYSTEM

VELOCITY / TIME PLOT
UPSTREAM OF CHECK VALVE FP2

GRAPH NO. 1.05

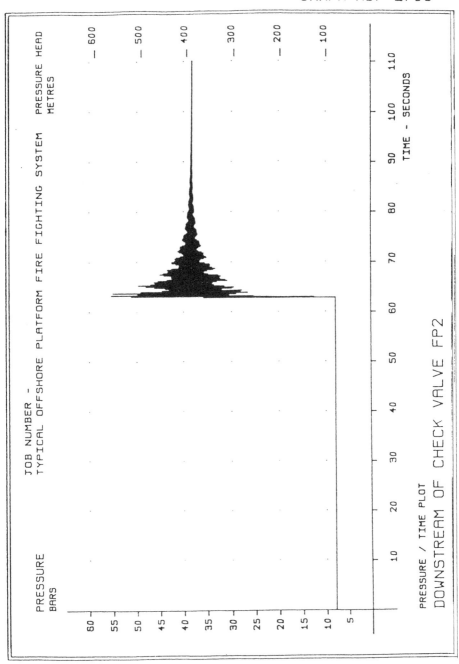

JOB NUMBER -
TYPICAL OFFSHORE PLATFORM FIRE FIGHTING SYSTEM

PRESSURE HEAD
METRES

PRESSURE
BARS

PRESSURE / TIME PLOT
DOWNSTREAM OF CHECK VALVE FP2

TIME - SECONDS

GRAPH NO. 1.06

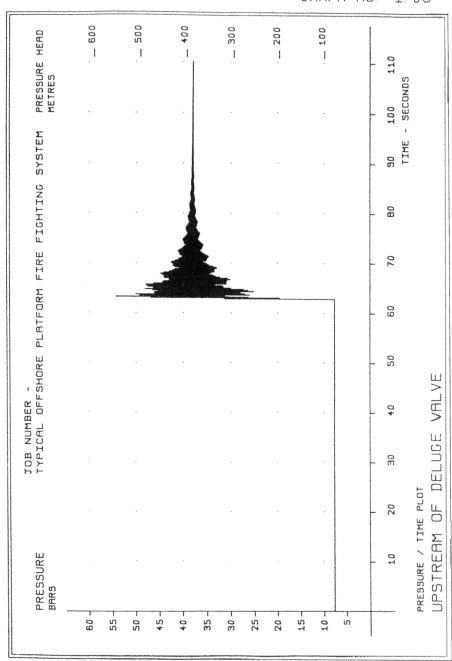

GRAPH NO. 1.07

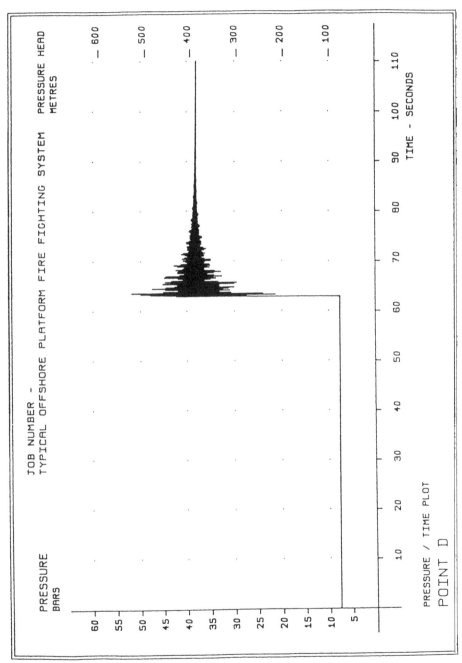

GRAPH NO. 2.01

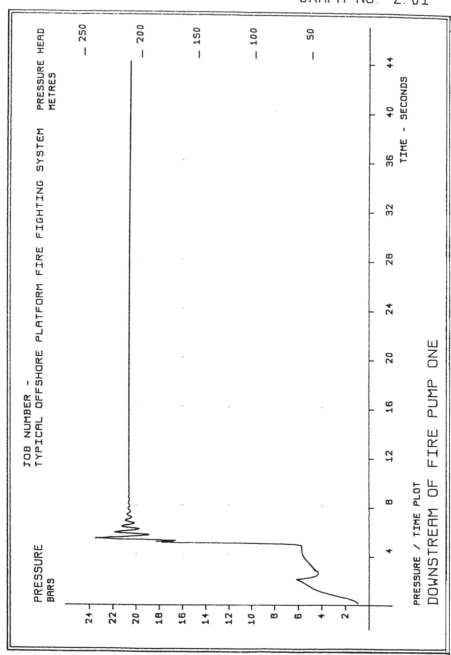

PRESSURE
BARS

JOB NUMBER -
TYPICAL OFFSHORE PLATFORM FIRE FIGHTING SYSTEM

PRESSURE HEAD
METRES

PRESSURE / TIME PLOT
DOWNSTREAM OF FIRE PUMP ONE

TIME - SECONDS

GRAPH NO. 2.02

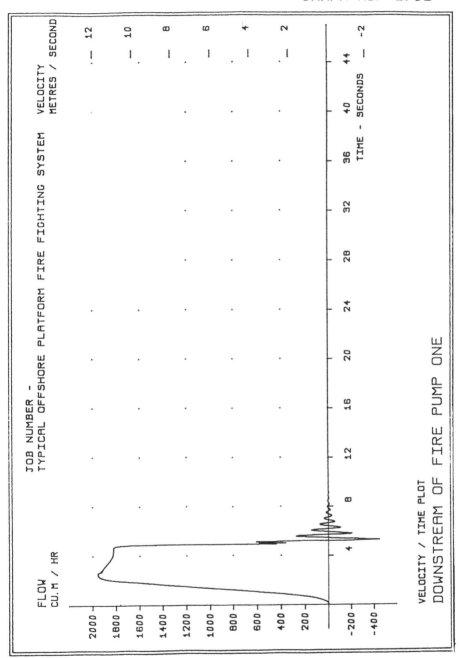

VELOCITY / TIME PLOT
DOWNSTREAM OF FIRE PUMP ONE

GRAPH NO. 2.03

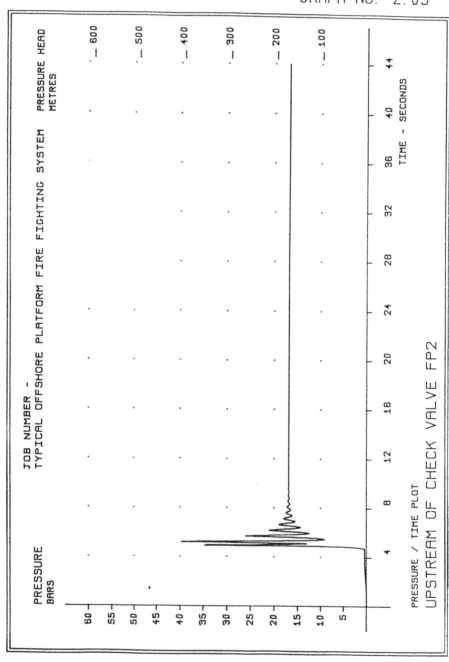

PRESSURE / TIME PLOT
UPSTREAM OF CHECK VALVE FP2

GRAPH NO. 2.04

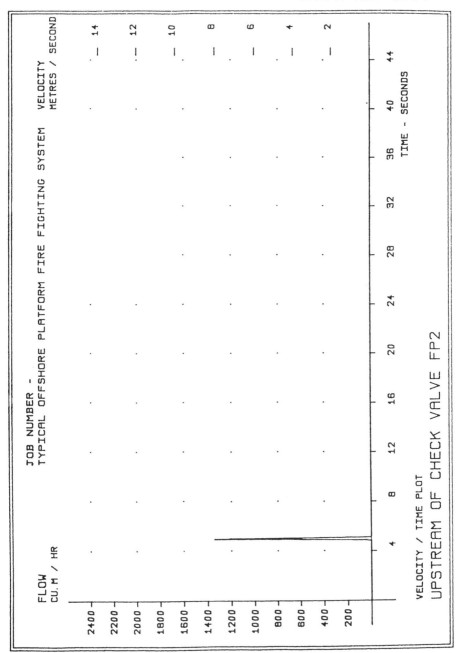

JOB NUMBER -
TYPICAL OFFSHORE PLATFORM FIRE FIGHTING SYSTEM

VELOCITY / TIME PLOT
UPSTREAM OF CHECK VALVE FP2

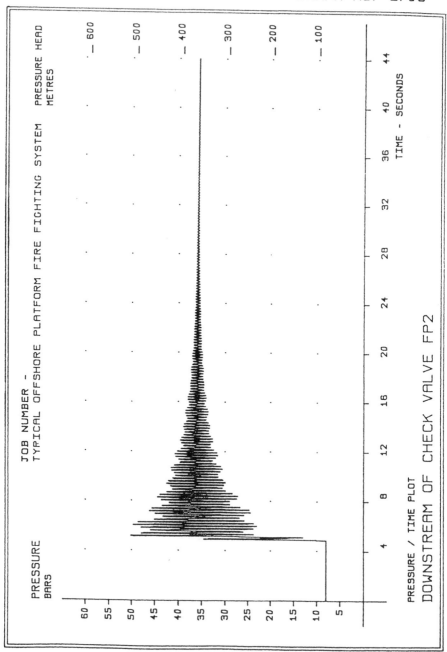

GRAPH NO. 2.05

GRAPH NO. 2.06

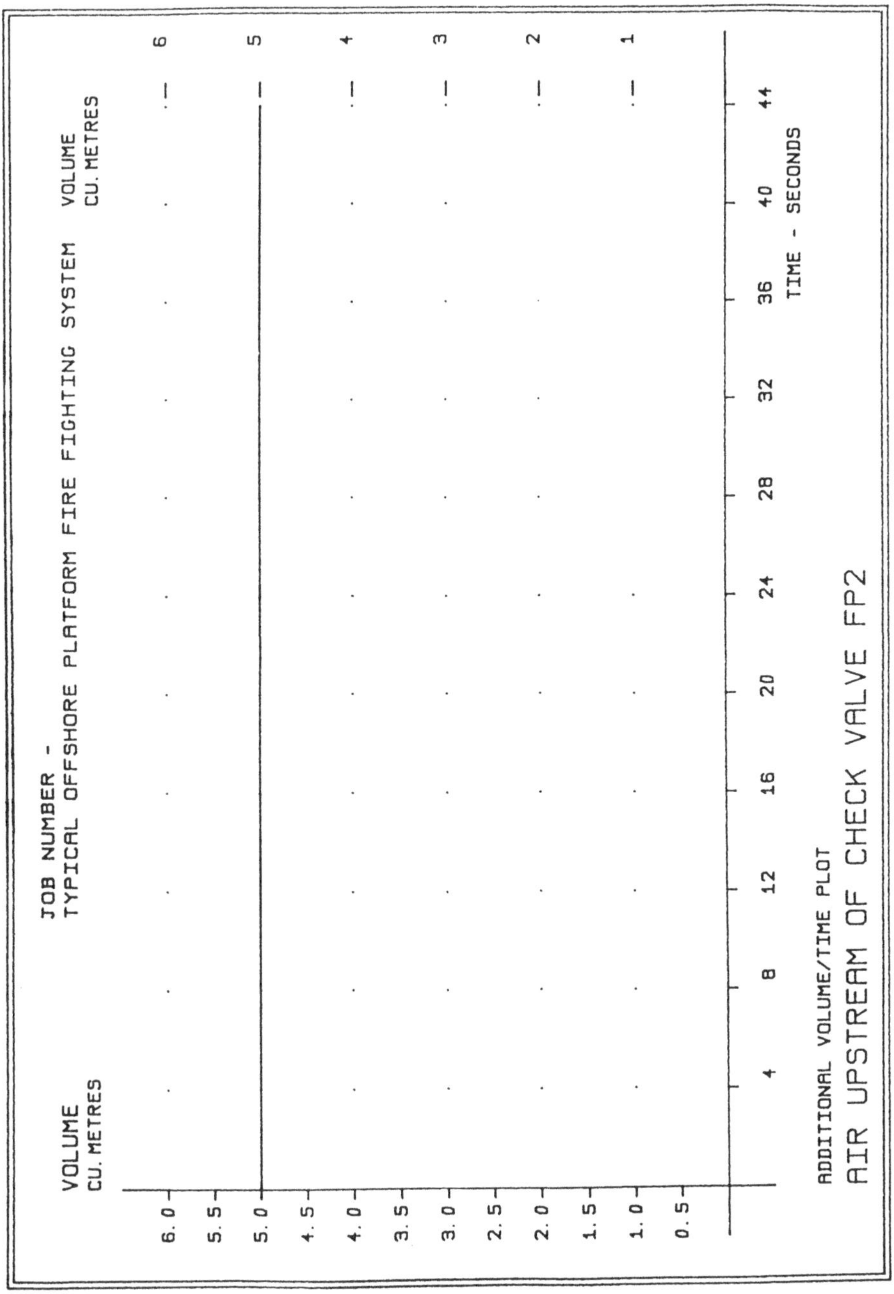

VOLUME
CU. METRES

JOB NUMBER -
TYPICAL OFFSHORE PLATFORM FIRE FIGHTING SYSTEM

VOLUME
CU. METRES

ADDITIONAL VOLUME/TIME PLOT
AIR UPSTREAM OF CHECK VALVE FP2

TIME - SECONDS

GRAPH NO. 2.07

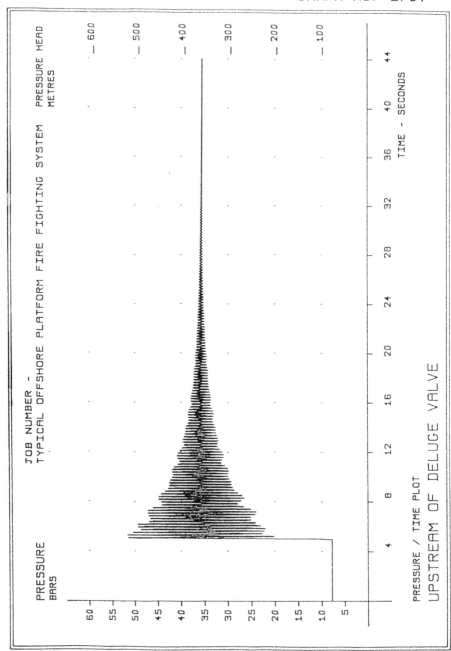

PRESSURE / TIME PLOT
UPSTREAM OF DELUGE VALVE

GRAPH NO. 2.08

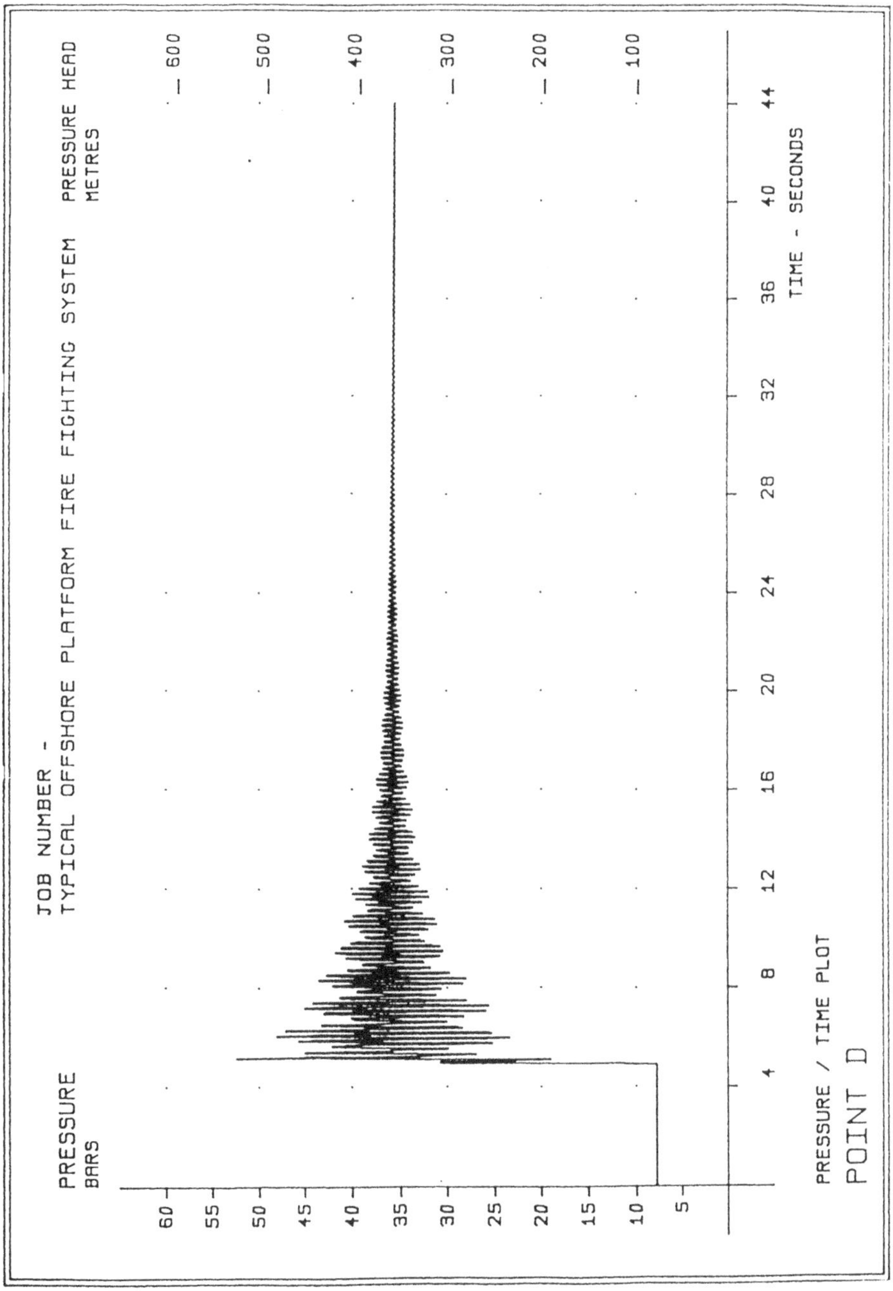

PRESSURE HEAD
METRES

— 600

— 500

— 400

— 300

— 200

— 100

JOB NUMBER -
TYPICAL OFFSHORE PLATFORM FIRE FIGHTING SYSTEM

PRESSURE
BARS

60
55
50
45
40
35
30
25
20
15
10
5

TIME - SECONDS

4   8   12   16   20   24   28   32   36   40   44

PRESSURE / TIME PLOT
POINT D

GRAPH NO. 2.09

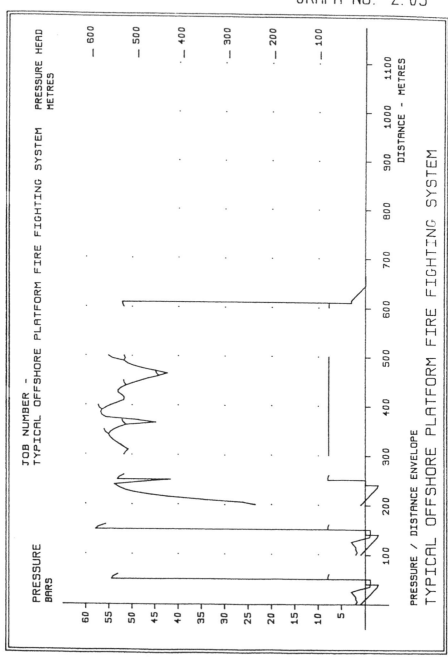

TYPICAL OFFSHORE PLATFORM FIRE FIGHTING SYSTEM

GRAPH NO. 3.01

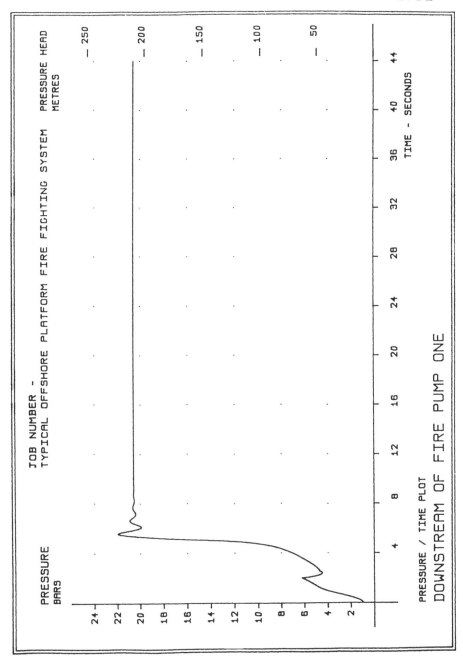

JOB NUMBER -
TYPICAL OFFSHORE PLATFORM FIRE FIGHTING SYSTEM

PRESSURE
BARS

PRESSURE HEAD
METRES

PRESSURE / TIME PLOT
DOWNSTREAM OF FIRE PUMP ONE

TIME - SECONDS

GRAPH NO. 3.02

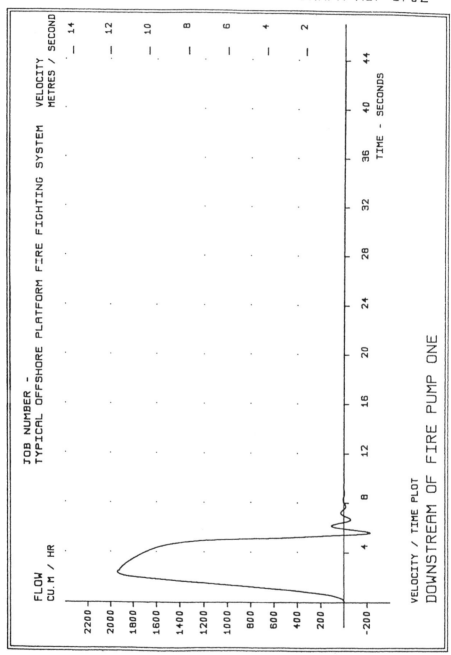

333

GRAPH NO. 3.03

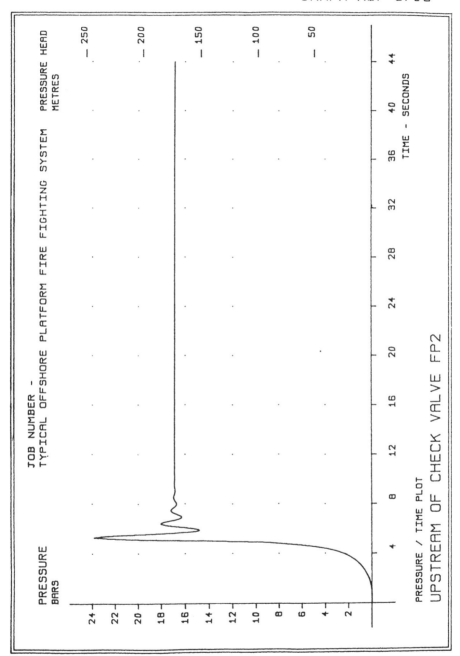

GRAPH NO. 3.04

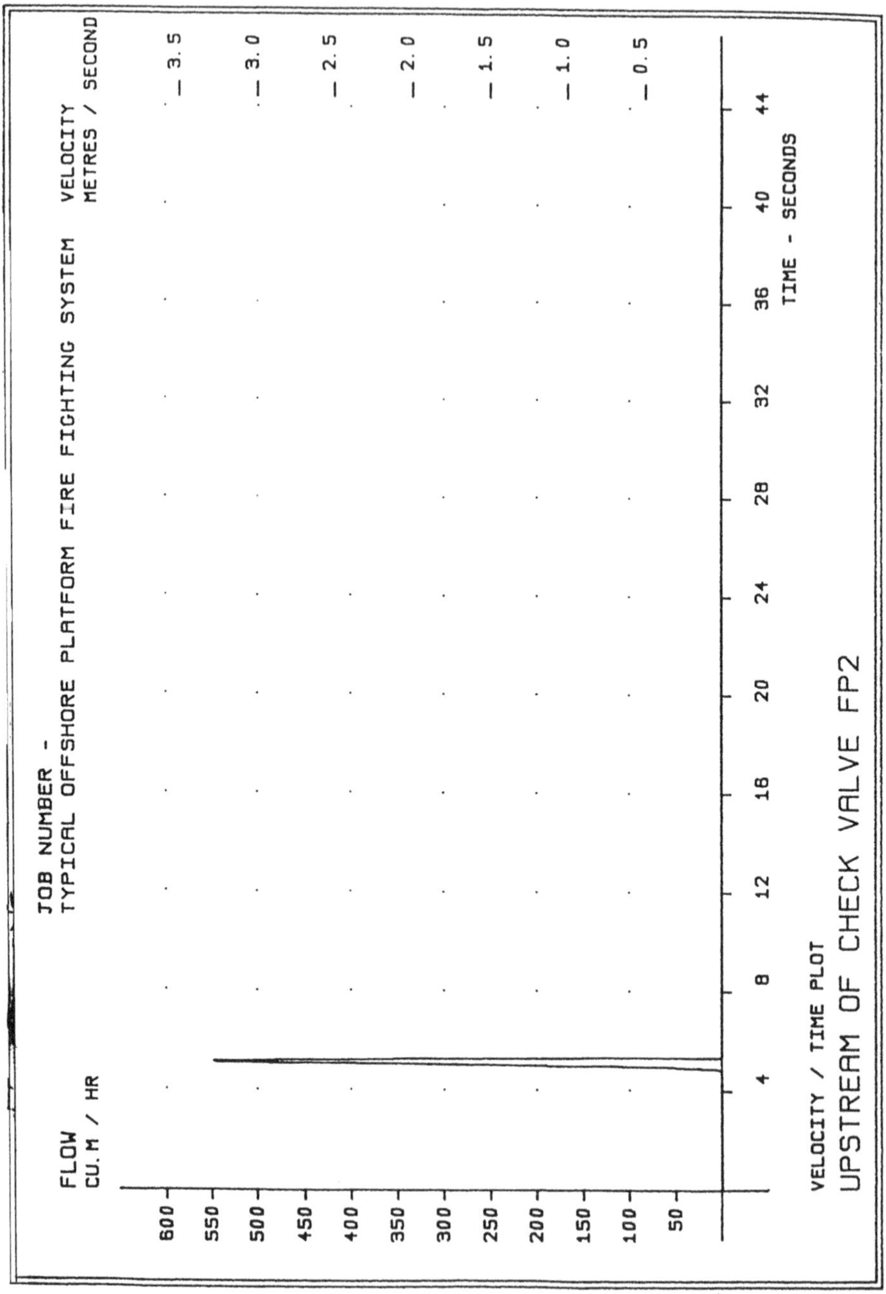
FLOW
CU.M / HR
JOB NUMBER -
TYPICAL OFFSHORE PLATFORM FIRE FIGHTING SYSTEM
VELOCITY
METRES / SECOND
600
550
500
450
400
350
300
250
200
150
100
50
3.5
3.0
2.5
2.0
1.5
1.0
0.5
4
8
12
16
20
24
28
32
36
40
44
TIME - SECONDS
VELOCITY / TIME PLOT
UPSTREAM OF CHECK VALVE FP2

GRAPH NO. 3.05

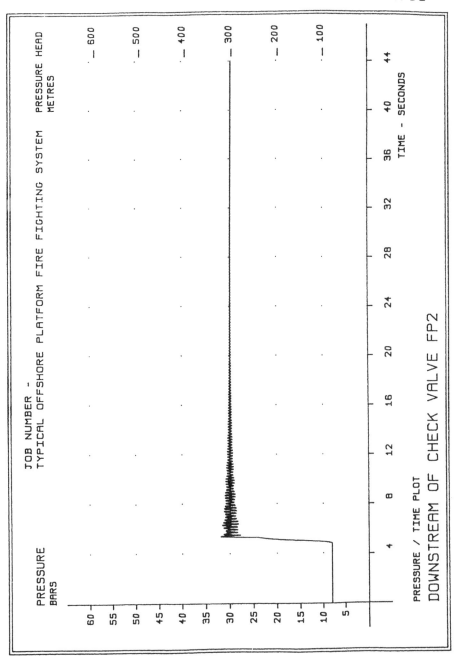

PRESSURE / TIME PLOT

DOWNSTREAM OF CHECK VALVE FP2

GRAPH NO. 3.06

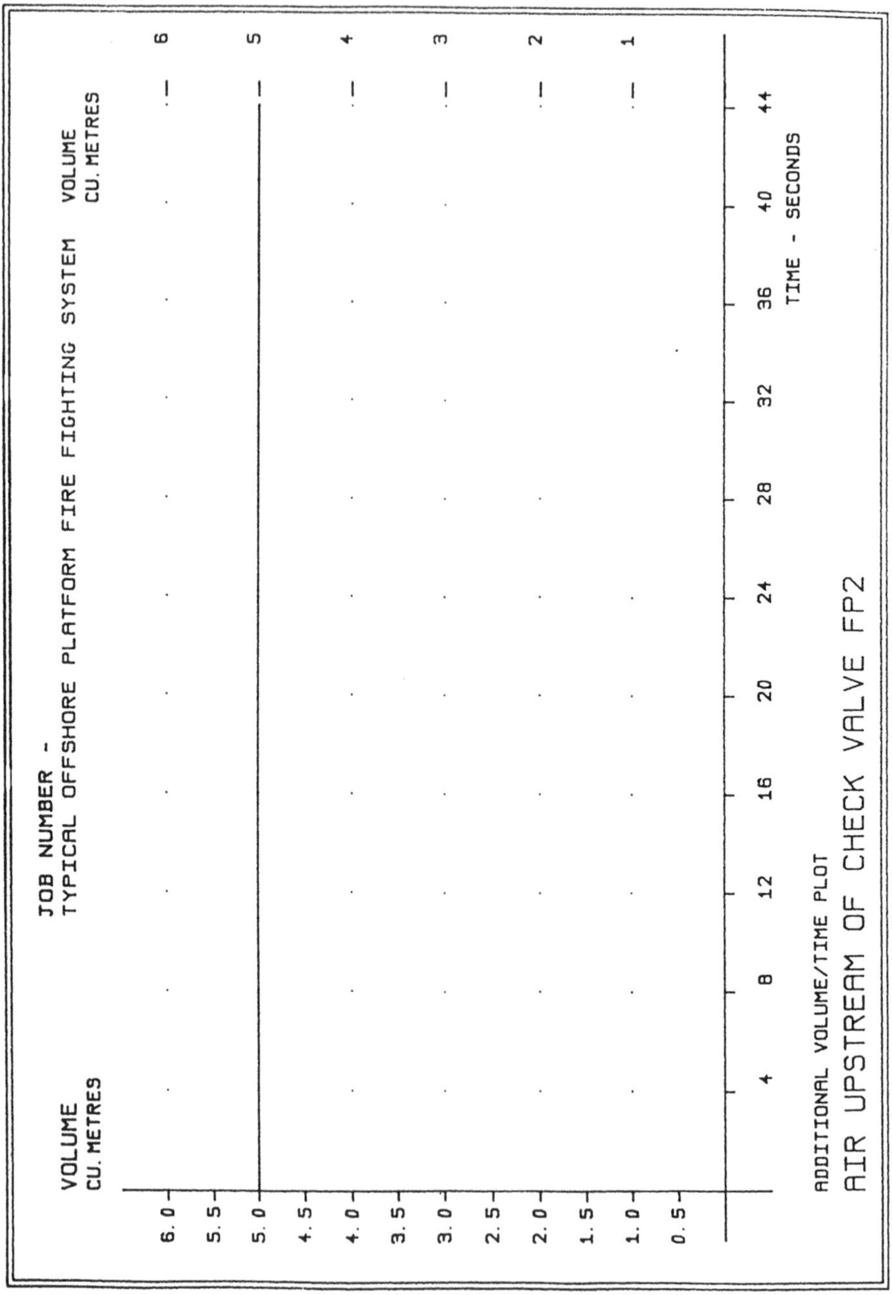

JOB NUMBER -
TYPICAL OFFSHORE PLATFORM FIRE FIGHTING SYSTEM

VOLUME
CU. METRES

VOLUME
CU. METRES

6 —
5 —
4 —
3 —
2 —
1 —

6.0
5.5
5.0
4.5
4.0
3.5
3.0
2.5
2.0
1.5
1.0
0.5

4   8   12   16   20   24   28   32   36   40   44

TIME - SECONDS

ADDITIONAL VOLUME/TIME PLOT

AIR UPSTREAM OF CHECK VALVE FP2

GRAPH NO. 3.07

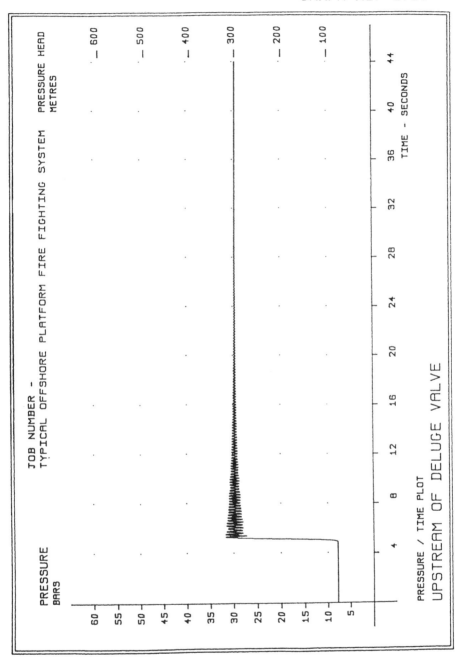

JOB NUMBER -
TYPICAL OFFSHORE PLATFORM FIRE FIGHTING SYSTEM

PRESSURE
BARS

PRESSURE HEAD
METRES

PRESSURE / TIME PLOT

UPSTREAM OF DELUGE VALVE

TIME - SECONDS

GRAPH NO. 3.08

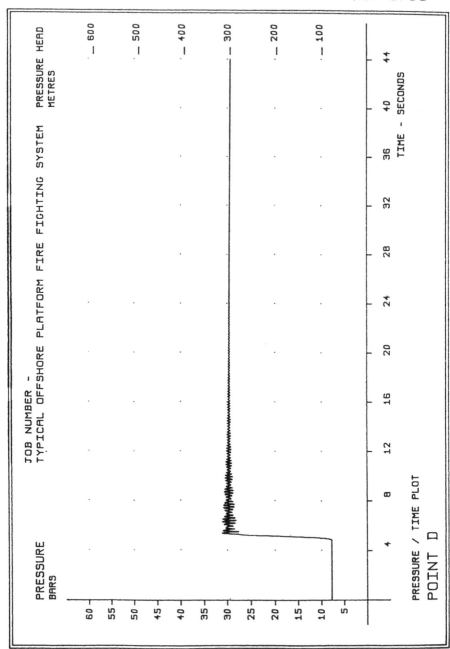

GRAPH NO. 3.09

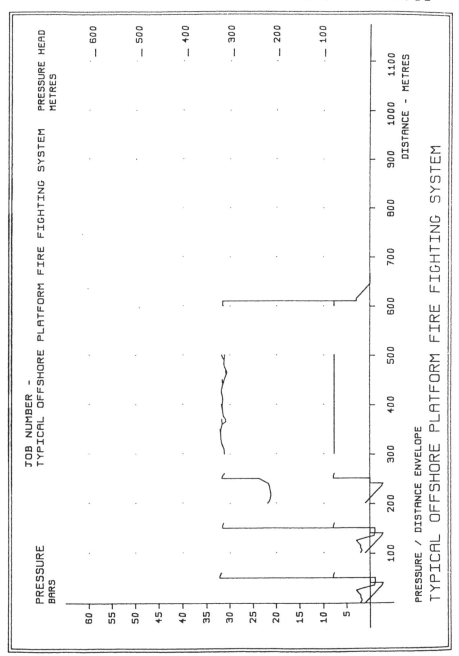

PRESSURE / DISTANCE ENVELOPE

TYPICAL OFFSHORE PLATFORM FIRE FIGHTING SYSTEM

HALON ALTERNATIVES AND REPLACEMENTS

By R A Whiteley, B. Tech, M.I. Mech. E., C. Eng.,
Chief Engineer, Wormald Manufacturing

Summary

The 1987 Montreal Protocol has led to a planned phase out of Halon
1301 by the year 2000. Halon's unique fire suppression properties of
low concentrations, low toxicity and no agent clean up requirements
has meant that no one existing fire fighting agent can replace it.
The applications and limitations of the various alternative agents
are reveiwed as well as the applications engineering for the
systems. The development status of the candidate replacements for
Halon are reported and the inhibiting factors appertaining to each
are discussed.

Introduction

Due to its unique properties Halon 1301 has become widely used as a
fire suppression agent for manned areas where a clean agent with low
toxicity is required. Due to the environmental damage caused by
Halon its reduction and elimination are being mandated for completion
by the year 2000. The other existing fire fighting agents and
systems have areas of application but also limitations. As a result,
careful hazard analysis will be required to determine the optimum
alternatives to Halon. Replacements for Halon are under development
by a number of chemical companies. Due to the stringent testing and
protracted time scales involved, any replacements will not be
available for some time and appear to have limitations of their own.

Halon 1301

Bromo trifluoromethane (B.T.M.) has been one of the most popular and
successful fire suppression agents in recent times. It has a unique
combination of properties which allows it to be used to protect Class
A hazards such as wood, paper, Class B hazards such as flammable
liquids, Class C hazards flammable gases, and Class E hazards - live
electrical equipment, all of which must be totally enclosed. Halon
1301 is:-

* Clean
* Odourless
* Colourless
* Tasteless
* Non-toxic
* Electrically non-conductive

By inhibiting the chemical chain reaction of combustion, it requires
concentration as low as 2.5% by volume in air in order to extinguish
a fire; however, as it provides no cooling effects so the Halon
concentration should be maintained until hot surfaces cool below
their auto ignition temperature.

As B.T.M. has a vapour pressure of 15 bars (200p.s.i.) at 20°C it
is stored in tanks as a liquid and super-pressurised with nitrogen to
either 25 bar or 42 bar.

Halon systems are designed to rapidly establish a uniform concentration of Halon in air (usually 5%) to ensure prompt extinguishment and minimum acidic products of decomposition. These products are produced when Halon 1301 comes into contact with flames and/or surfaces at temperatures above 475°C. Plant and equipment in the hazard area are normally shut down prior to Halon release, as is the HVAC system followed by closure of doors and fire dampers in ductwork.

The Ozone Issue

The use of Halon 1301 has been growing steadily at approximately 10% p.a. with 22% p.a. being released and the balance remaining in tanks. Of the amount released, the largest single useage had been for discharge testing of systems upon commissioning. This had become accepted practice due to the unacceptably high failure rate of systems when tested. Figures as high as 70% failure rate have been quoted for Halon systems tested, due primarily to failure of the enclosure to retain Halon adequately. (Reference figures 1, 2 and 3).

The general test criteria is for the hazard to retain 80% of the initial concentration for 10 minutes at the height of the highest risk within the hazard, (Ref. BS5306 part 5.1).

Through the 1980's the British Antartic Survey had detected a rapid deterioration in ozone levels above the Antartic within the winter vortex. The Earth's ozone layer filters out the cancer causing ultra violet rays from the sun.

The cause of these depletions of up to 50% was traced to a family of man-made chemicals used in large quantities over the past 30 years. These are known as chlorofluorocarbons (C.F.C.'s), used principally in refrigeration, aerosols, and the manufacture of expanded foam packaging. Whilst relatively small quantities were used as Halon 1301 and Halon 1211 (B.C.F. - used in portable extinguishers) these latter C.F.C.'s were considered to be up to 10 times more damaging per kg than other C.F.C.'s. When their Ozone Depletion Potential (O.D.P.) is taken into account, the Halon are deemed to account for 14% of the damage. (Reference figures 4 and 5).

As a result of the scientific evidence, Governments met in 1987 and signed the Montreal Protocol to phase out C.F.C.'s. The Protocol had one section for commercial C.F.C.'s and a separate one for the unique fire suppressant Halons.

The Protocol, as currently updated, and adopted by most countries in the World, calls for a 1992 cutback in Halon production to 1986 levels, followed in 1995 by a further 50% cut leading to a 'phase out of Halons by the year 2000'. This time-table is being written into regulations such as the International Maritime Organisation, who propose no new Halon systems after 1992 and replacement of all existing Halon systems by 2000. Similar regulations are proposed for the U.K. North Sea Offshore Installations.

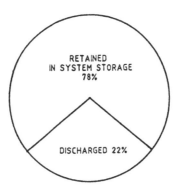

fig. 1 DISTRIBUTION OF ANNUAL HALON 1301 & 1211 PRODUCTION (USA 1985)

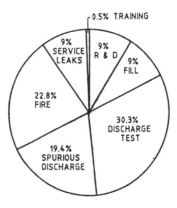

fig. 2   ANALYSIS OF HALON 1301 RELEASES (USA 1985)

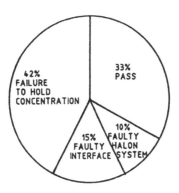

fig. 3   ANALYSIS OF HALON DISCHARGE TESTS (IRI 1979)

Substance	Domestic Use	Ozone Depletion	
		Factor*	% of Total
CFC-11	75.0	1.00	23
CFC-12	135.0	.86	35
CFC-113	63.0	.80	15
Halon 1211	2.8	2.39	2
Halon 1301	3.6	11.43	12
Methyl chloroform	292.0	.15	13

* These weighting factors are based on preliminary analysis using 1-D parameterization (Wuebbles) and are subject to revision.

Data on domestic use of CFCs from Rand and Radian. Data on domestic use of halons from *Industrial Economics* based on industry surveys.

**1985 US Production and Domestic Use of
Fully Halogenated Substances and Methyl Chloroform
(in thousand metric tons)**

FIGURE 4

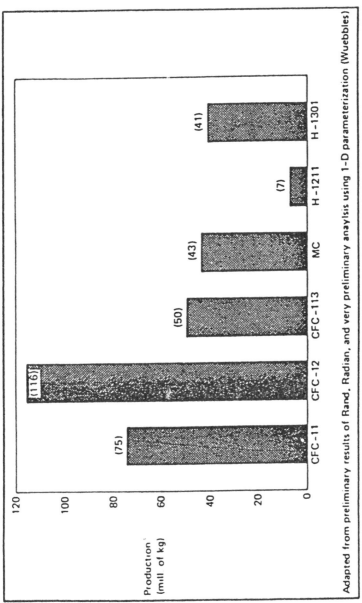

Adapted from preliminary results of Rand, Radian, and very preliminary anaylsis using 1-D parameterization (Wuebbles)

US use of fully halogenated substances and methyl chloroform in 1985, weighted.

FIGURE 5

Halon 1301 is used to protect a wide range of hazards. Whilst some may be relatively low fire hazards, many are vital to the continued operation of offshore facility. Typical hazards include:

Computer Rooms	Radio Rooms
E.D.P.	Gas Turbine Enclosures
Control Rooms	Pump Rooms
Data Storage	Flare/Vents
U.P.S.	Switchgear
Battery Rooms	

Due to the failures of systems to maintain Halon concentrations during tests, it had been common practice to discharge Halon as part of commissioning to verify the system and, more importantly, the integrity of the enclosure. This latter problem is now effectively addressed by means of an air fan integrity test as per the latest draft of BS 5306 part 5. This enables the wide Halon descending interface to be modelled using a portable test kit and lap top computer.

The alternatives to Halon:-

There are a number of alternative agents which have advantages and disadvantages when considered as alternatives to Halon for each application. Water, foam, Carbon Dioxide and Dry Powder, all can be considered but no one agent can readily replace the properties Halon provides. Each will be considered in turn.

CO2

Is the only clean agent requiring no clean up after discharge. It is odourless, colourless, electrically non-conductive, but suppresses fire by dilution of oxygen and so is lethal to anyone in or entering a hazard containing CO_2. For this reason, CO_2 systems may not be automatically actuated for normally manned areas and must be set for manual operation when unmanned areas are entered. Relatively large quantities of CO_2 are needed to dilute the air, with 34% CO_2 by volume being the minimum requirement. As a result, large quantities of high pressure cylinders must be accommodated for each risk. The cylinders are on long (9-12 months) lead time from manufacturers, thereby creating supply/stock problems for the fire protection companies. CO_2 is suitable for all types of fires and is readily available.

Sprinklers

Are relatively slow response heat detectors which release water over the seat of a fire when activated. Water will control and extinguish fires in ordinary combustible materials; it will only cool flammable liquid and gas fires, and is unsuitable for fires in live electrical equipment. Once electrical equipment has been shut down, however, it becomes ordinary combustibles.

Most modern electronic equipment is solid state and is unaffected by water. The equipment can be dried out and put back in service. Sprinkler protection for computer suites is advocated by such noteable organisations as Factory Mutual.

Pre-Action Sprinklers

Are used to avoid the risk of water from damaged or leaking systems affecting sensitive electrical/electronic equipment. The pipework in a pre-action sprinkler system is charged and supervised for leaks by air. Only when sensitive smoke (or other) detectors are activated is water released into the system so that it is primed ready should the fire develop sufficiently to activate one or more sprinklers. Pre-action sprinkler systems are an accepted alternative to Halon protection of computer/E.D.P. facilities.

Foam is available as low, medium or high expansion. Low expansion (<20:1) is effective on flammable liquid fires by cooling and separating air from the fuel. Medium and high expansion foam may be used on flammable liquids and on ordinary combustibles. They use relatively small quantities of water but are electrically conductive, disorienting to people within the foam, and may be unable to flow round/through congested hazards such as cable ways.

Waterspray

Uses either medium or high velocity nozzles for zone deluge usually to protect flammable liquid hazards. It maintains the integrity of vessels and structures by cooling and achieves fire suppression by emulsification and steam smothering. Drainage requirements must be taken into consideration to avoid spread of flammable liquids on the water.

Dry Powder

Dry Powder is effective on all types of fire by ingestion of powder into the combustion air stream. Whilst it achieves rapid extinguishment it provides no cooling and the widely dispersed fire powder can present serious clean up problems.

Detection

Detection only may be an option using high sensitivity detectors to shutdown plant whilst in the early, smouldering phase, of fire. Such systems rely upon shutdown before the fire becomes self propagating and assuming manual intervention by responsible personnel.

As is evident from the above, no single agent can readily replace Halon 1301 although the various agents can be used for the fire hazards protected by B.T.M. A matrix showing preferred options for various hazards is given in Figure 6.

The replacements for Halon

Several chemical companies, who currently produce Halon, have been researching alternative agents. The Companies known to be active in this area are Great Lakes Chemical Company, Du Pont, I.C.I., and Autochem. To date all have announced that they have candidate agents in development and all have indicated that the environmental and toxicological testing by the regulatory authorities will take several years. The current situation is as follows:-

FIGURE 6 - HALON ALTERNATIVES MATRIX

Typical	MV Spray	HV Spray	Carbon Dioxide	Dry Powder	Low Ex Foam	Med Ex Foam	Hi Ex Foam	Spklrs Std	Spklrs Quick Response	Spklrs On/Off	Pre-Action Spklrs
Actual Computer Room			3 M.I.					2	1	2	2
Under floor			1					2	2	2	2
Equipment Cabinets			1 I								
Tape Store			3 M.I.					2	1	2	2
Printer Room			3 M.I.					2	1	2	2
U.P.S.			1 I	2				3	3	4	
Transformer Rooms		1	2 M.I.	3	4	4	4	5	5		
Switchgear Rooms			1 I	2				4	4		
Diesel Gen.	2		1 I	3	4	5	5	4	4		
Flamm. Liquid Stores	3	3	1 I	2	4	5	6				
Control Rooms			1 M.I.								1
Cable Rooms	1		3 I	4		3	2	2			
Gas Turbines			1 I	2							
Telecoms Room			4 I					3	2	2	1
Battery Room			1 I								
Product Pump Rooms	1		2 I	3	4	5	6				
Engine/Boiler Rooms	2		1 M		3	5	4				
Dust Explosion Risks				1							
Flares/Vents				1							

1, 2, 3, 4, 5, 6 - Preferences, M.I. - Manual or with Interlocks, M - Manual, I - With Interlocks

G.L.C.C. appears to be the most advanced having received preliminary approval from the American Environmental Protection Agency (E.P.A.) to begin pilot production of 'Firemaster 100', samples of which are now available. While it is as an efficient fire suppression agent as Halon, its toxicity is much greater than Halon 1301. This, and its low vapour pressure would suggest its most likely application would be as a replacement for Halon 1211. (Reference figure 7).

Two questions have been raised with regard to 'Firemaster 100'. The first is that it is claimed during initial tests to be mutagenic to bacteria. However, the test basis is being called into question and the outcome is unknown. The second point concerns its O.D.P. (Ozone Depletion Potential). Initial anouncements quoted an O.D.P. of 0.5, which others claim will prove nearer 0.7. Later predictions suggest its O.D.P. will meet the latest U.S. Clean Air Act requiring an O.D.P. no greater than 0.2.

G.L.C.C. also say they have a B.T.M. alternative, with zero O.D.P. in its early stages of development.

Du Pont have announed candidate agents to replace both 1211 and 1301, namely FE232 and FE25 respectively. FE232 is claimed an O.D.P. one twelfth of Halon 1211 (2.3) but is believed to be three times more toxic than 1211 and half as efficient, i.e. twice as much agent would be required compared with 1211. FE25 is claimed to have zero O.D.P. No toxicity characteristics are known, but again, it is only half as efficient as 1301 and likely to cost two to three times the price per kg.

I.C.I. announced in October, a candidate replacement for B.C.F., namely 124B1 with similar density, boiling point and critical temperature to Halon 1211. Toxicity data suggests it would be favourable but its O.D.P. may be above the 0.2 now mandated by the U.S. Clean Air Act.

Atochem have also stated they have identified candidiate agents but have, as yet, not released any more precise information. It is, however, claimed to have 0 O.D.P., 10 bar vapour pressure and requires double the extinguishing concentration of Halon 1301.

E.P.A. toxicity testing takes several years and all companies talk in terms of 1995+ before a replacement for Halon 1301 is likely to be approved/released. It is clear that the companies are continuing to expand R & D and approval expenditure which suggests that a replacement with similar characteristics to Halon 1301 may well evolve in time.

FireMaster™ 100

PROPERTY COMPARISON

Patent Pending

	FireMaster™ 100	HALON 1211	HALON 1301
Chemical Structure	CHF_2Br	CF_2ClBr	CF_3Br
Molecular Weight	130.9	165.4	148.5
Boiling Point °C °F	 -15 5	 -4 25	 -58 -72
Density @ 70°F lb/gal lb/ft^3	 12.9 97	 15.2 114	 13.1 98
Electrical Conductivity	Non Conductive	Non Conductive	Non Conductive
Vapor Pressure (psia) @ 20 °F @ 70 °F @ 120 °F	 22 59 136	 13 37 79	 100 214 401
LC_{50} 4 hr. Acute Toxicity (ppm)	108,000	131,000	>800,000
Extinguishing Concentration, N-Heptane (mg/l),(% vol)	214 4.0	229 3.8	189 3.5
Ozone Depletion Potential (ODP)	0.5 (estimated)	3.0	10.0

he information supplied is presented in good faith and has been derived from sources believed to be reliable. Since conditions of use are beyond our control, all risks are assume y the user. No representation is expressed or implied, and nothing herein shall be construed as permission or recommendation to practice a patented invention without a license

FIGURE 7

Conclusions

The fire protection industry saw a significant drop in demand for Halon 1301 when the ozone issue arose. More recently, the demand for Halon has stabilised as, it would appear, the more essential uses for Halon are being identified and maintained. A wide variety of alternatives are being used to replace Halon with pre-action sprinklers and CO_2 being perhaps the more common. As part of the industry wide re-evaluation of the use of Halon 1301 it appears that the use of modular Halon systems is finding favour. The philosophy being to only put Halon into the relatively small spaces where the actual fire risk exists thereby using small Halon tanks activated by detection specific to the particular cabinet or enclosure. This has become a practical proposition using addressable detectors and actuators.

The market for alternative agents to Halon 1301 will undoubtedly be much smaller than that which was enjoyed by B.T.M., due to the re-assessment of hazards and the acceptance of various alternatives. The likely high cost will also be a factor but this will be minimised by the increased use by fire engineers of modular Halon/alternative agent systems.

There has proved to be no simple, or ready fire safety solution to the demise of Halon but the careful review of hazards and their protection needs by Fire Engineers will lead to optimisation of protection with an eye towards technological developments in the future.

THE EVALUATION AND TESTING OF FIREWATER DELUGE SYSTEMS

By W Fitzpatrick
Wormald Engineering
and R A Whiteley, B. Tech, M.I. Mech.E, C. Eng
Wormald Manufacturing

Summary

Due to the various changes made in process areas and firewater deluge systems, it has been proved highly desirable for experienced fire systems engineers to carry out a detailed post construction evaluation/audit. Such audits have particular benefits detecting deficiencies in nozzle placement and nozzle coverage, leading to full computer modelling of the entire firewater systems including ring main loops and pumps against all operating scenarios.

The paper will highlight the steps taken in these evaluations and the problems encountered and benefits derived. It will also discuss operational experience associated with deluge testing such as nozzle blockage, valves, drainage and coverage assessments.

Introduction

Operational experience has revealed an increasing need for careful
and comprehensive evaluation of firewater deluge systems and water
supply networks post construction. Of necessity this work must be
carried out by fire engineers with firsthand experience of deluge
system design, performance and evaluation. The identification of
need and selection of a qualified response enables operators and
certification authorities to ensure that vital firewater safety
systems are operational and meet the required performance parameters.

Deluge Systems - the problems

Problems can occur at several stages during the design and
construction phases of a project. During the design phase, the size
and positioning of pipes, ducts, cable trays and vessels is changed
with each change needing to be accommodated by the firewater deluge
system. Problems arise when changes are made late in the design
phase and during construction. Site problems can arise due to some
services not being installed to drawing thereby creating a knock on
effect into other services and affecting the satisfactory coverage of
the deluge system. The practice of leaving deluge pipework below
80mm to be site run can also lead to subsequent problems. The deluge
pipework below 80mm typically represents 60% of the total, which is
site run over and around the various obstacles to reach the
pre-selected nozzle positions. Little, if any, consideration is
given to the inhibiting effects of addition pipe fittings and pipe
runs on the system hydraulics computed to ensure that the required
densities can be achieved for fire containment. In other instances,
the effects of additional piping and additional nozzles impacts upon
the ability of the limited pumping capacity to fulfil the resultant
enhance requirements. Surveys have found pipework changes which
require 30, 40 and 50 bar pump pressures before systems would be
effective.

Other situations arise where due to proximity of large pipes/ducts,
relocation of nozzles, or lack of provision for numerous small
services, the necessary water distribution can be dangerously
disrupted leaving underprotected zones where hot spots can develop
which could lead to critical loss of containment.

In older deluge systems, additional problems are encountered. One is nozzle clogging by corrosion within the pipe, large particles in suspension and/or construction debris. This situation can be exacerbated by use of non-proprietary small orifice nozzles which require high quality water supplies for satisfactory performance. Another common occurrence is where services are added or modified due to operational needs, but without recourse to fire engineers to ensure the appropriate upgrading of the protective systems. The deterioration of systems and deluge valves has also been found, whereby systems have become inoperable as a result of insufficient maintenance. Modifications to existing systems can, when carried out by non-specialist personnel, result in overloading of water supplies with the consequence that critical areas become starved of the minimum density coverage they require.

Analysis

Problems of the various types as described above can only be identified through careful and thorough analysis by fire engineers with the knowledge of nozzle performance and limitations, discharge characteristics, system hydraulics also able to carry out computer hydraulics modelling of complete firewater system scenarious back up by actual system performance testing of flow densities and running pressures. Such analyses require a rigorous methodology based on sound fire engineering practice and applicable to any deluge system regardless of design, age or condition. The overall objectives of a fire water system audit would typically be:-

1. Establish the existing firewater system design philosophy.

2. Verify the detailed 'as built' status of the firewater systems.

3. Identify any design shortcomings of the systems.

4. Establish by testing the performance shortcomings.

5. Investigate the optimum means of bringing the systems up to meet their specified operational criteria.

6. Prepare detailed work packs to carry out, test and commission the rectified systems.

7. Produce final 'as built' certification and documentation.

The existing firewater design/philosophy may be used to establish preliminary system reference data. Flow diagrams, plus piping and instrument diagrams, would be required along with the method(s) of deluge actuation.

The sequential response for detection through to operation should be established and must include pump start up and deluge valve release. The number of fire pumps and the sequential start arrangements must also be pre-determined. The system 'General Arrangement' drawings and the individual firepump performance characteristics are needed as input data for computerised hydraulic analysis.

Current 'as built' status may be determined by a survey of the
systems against the latest drawings with careful recording of
discrepancies. The visual examination by qualified fire engineers
will provide information covering a number of key areas.
Unrecorded changes in the size and routing of pipes, nozzles and
obstructions affecting deluge performance can be highlighted.
Preliminary identification can be made of inadequate coverage for
existing or additional plant, plus the visual examination of the
pipework nozzles and valves will establish their overall condition.

Once accurate information has been collated on the deluge systems
pipework and nozzles, firemain, and fire pump characteristics, this
information can be fed as basic data into a computer model of the
entire facility firewater systems including looped mains and multiple
firepump locations. This then enables any fire scenario to be
hydraulically modelled and the resultant output used to determine
potential shortcomings in the overall protection. These may
include:-

Insufficient flow and/or pressure to the nozzles, overloading of
pumpsets, and excessive flows and flow losses in critical sections of
pipe. The nature and location of such deficiencies can be expected
to vary according to the various operational scenarios being
modelled. The speed and flexibility of such computer models also
enables engineers and management to examine 'what if...' situations
quickly, serve to address particular concerns, and to evaluate
various options for remedial courses of action.

Before finalisation of the design review and computer models, it is
virtually important to establish by testing the performance
shortcomings. Such tests measure pressures, flows, densities and
nozzle performance using as reference points the fire pumps,
firemain, deluge control valves, most hydraulically remote nozzles
plus the hydraulically most favourable nozzles. Test equipment used
must be capable of accurate measurement but must, at the same time,
not be susceptible to damage or disruption from foreign matter which
can be present within the systems being tested. Such equipment would
include:-

Pre-calibrated pressure gauges.
Flow socks.
Collecting receptacles.
Density measurement trays.
Waterproof stop watches.

Of necessity, such test programmes need to be carefully planned and
co-ordinated with the data being collated into a test pack for
presentation, investigation and record purposes. These test packs
include the original design philosophy/specification, flow sheets,
P & I Ds., Firemain 'as built' G.A.'s or isometrics, fire pump
performance curves, existing hydraulic calcs and noding isometrics,
equipment literature, and equipment layout/plot plans.

Testing

The results of actual system tests may be used to verify the design and computer analysis but also highlight additional problems due to the age/condition of the pipework, design limitations, equipment quality as well as the adequacy (or otherwise) of existing maintenance procedures. By comparing the actual performance data against the design predictions, it is normally possible to identify, with a high degree of accuracy, where deficiencies have occurred.

Blockages are a major problem, particularly in old and/or poorly designed systems. Internal corrosion caused by the saline environment results in large and small particles becoming lodged within sections of pipework and in nozzles causing reduced water flow/coverage to critical plant and hazard areas.

Also of major concern has been the condition of pumps, and both isolating and control valves which suffer from ageing, poor design and, in some instances, poor maintenance. Instances are not uncommon where control valves fail to operate when tested, which would in an emergency situation, have catastrophic results. The fact that such conditions are often only highlighted during assessment of deluge systems raises serious questions for operations management.

Practical testing is a vitally necessary complement to the design analysis as it establishes the actual efficiency and operational readiness of the existing systems.

Remedial Action

The analysis of the results from the design audit and testing require careful evaluation before an appropriate remedial programme can be drawn up. The first steps are always to identify and locate problem areas and shortcomings. Whilst some remedial actions may be self evident, many require further evaluation of various options in order to determine optimum solutions which satisfy the regulatory, technical, financial and operational requirements. The remedial action phase of these type of studies demands a high level of planning and co-ordination. This becomes necessary to synchronise the re-design of pipework and its procurement, fabrication and supply; the selection and procurement of replacement valves and nozzles, co-ordination of construction manpower with permits to work to erect new pipework and to clean clogged nozzles and pipework. An integral part of this phase will be the upgrading of maintenance procedures and the training of personnel.

Upon completion of the remedial work on various sytems, they must be tested and commissioned to verify the satisfactory performance of the systems as amended.

Records

As with all jobs, they are not complete until the paperwork is done.
Accurate records of the new 'as built' systems are needed for future
reference plus the test certification and results documentation.
These can then be made available for inspection by the certifying
authorities and serve as valuable accurate records for engineers when
future changes and upgrades to services and facilities take place.

Conclusions

The analysis and auditing of total firewater systems requires an
indepth knowledge and understanding of design parameters, regulatory
requirements, equipment performance and limitations, hydraulics plus
the practical experience of system testing and trouble shooting.
Only when such expertise is deployed in system evaluations can fire
protection performance be established to the satisfaction of
authorities, management and facilities personnel.

Acknowledgements

The authors wish to acknowledge the contributions to this paper made
by Wormald personnel who have shared their expertise and experiences
in ensuring this paper is an accurate reflection of system audits.

.